Language in mathematical education

Language in mathematical education

Research and practice

Edited by Kevin Durkin and Beatrice Shire

Open University Press
Milton Keynes · Philadelphia

Open University Press
Celtic Court
22 Ballmoor
Buckingham MK18 1XW

and
1900 Frost Road, Suite 101
Bristol, PA 19007, USA

First published 1991
Reprinted 1993, 1995

British Library Cataloguing in Publication Data

Language in mathematical education.
1. Education. Curriculum subjects: Mathematics. Linguistic aspects
I. Durkin, Kevin, *1950–* II. Shire, Beatrice, *1947–* .
510.7

ISBN 0-335-09367-1
ISBN 0-335-09366-3 (pbk)

Library of Congress Cataloging-in-Publication Data

Language in mathematical education: research and practice/edited by Kevin Durkin and Beatrice Shire.
p. cm.
Includes bibliographical references.
ISBN 0-335-09367-1 ISBN 0-335-09366-3 (pbk.)
1. Mathematics—Study and teaching. 2. Language and education.
I. Durkin, Kevin, *1950–* . II. Shire, Beatrice, *1947–* .
QA11.L377 1990
510'.071—dc20 90–32285 CIP

Typeset by Vision Typesetting, Manchester
Printed and bound in Great Britain by
Biddles Ltd, Guildford and King's Lynn

Contents

List of contributors

Julie Anghileri, Homerton College, Cambridge.
Jeffrey Barham, Department of Education, University of Cambridge.
Alan Bishop, Department of Education, University of Cambridge.
Philip C. Clarkson, Institute of Catholic Education, Victoria, Australia.
Richard Cowan, Institute of Education, University of London.
Alyson Davis, Department of Psychology, University of Surrey.
Erik De Corte, Centre for Instructional Psychology, The University of Leuven, Belgium.
Chris Donlan, Curriculum Development and Resource Unit, Invalid Children's Aid Nationwide, London.
Kevin Durkin, Department of Psychology, The University of Western Australia.
Karen C. Fuson, School of Education, Northwestern University, USA.
Lulu Healy, Institute of Education, University of London.
Celia Hoyles, Institute of Education, University of London.
Ella Hutt, Curriculum Development and Resource Unit, Invalid Children's Aid Nationwide, London.
Judith Kearins, Department of Psychology, The University of Western Australia.
Daphne Kerslake, Department of Education, Bristol Polytechnic.
Kazuhiro Kuriyama, Miyazaki Women's College, Miyazaki, Japan.
Youngshim Kwon, School of Education, Northwestern University, USA.
Pearla Nesher, University of Haifa, Israel.
Rita Nolder, Department of Education, University of Loughborough.
David Pimm, Centre for Mathematics Education, Open University.
Susan Pirie, Mathematics Education Research Centre, University of Oxford.
Beatrice Shire, Institute of Social and Applied Psychology, University of Kent at Canterbury.
Anne Sinclair, Faculté de Psychologie et des Sciences de l'Éducation, University of Geneva.
Rosamund Sutherland, Institute of Education, University of London.

Eva Teubal, University of Haifa, Israel.
Michael Thomson, East Court School, Ramsgate.
Lieven Verschaffel, Centre for Instructional Psychology, University of Leuven, Belgium.
Hajime Yoshida, Department of Psychology, Miyazaki University, Japan.

Section 1

Introduction

1 Language in mathematical education: An introduction

Kevin Durkin

Mathematics poses many challenges to pupils and their teachers. Why focus on language? The mathematician, after all, works in an abstract and highly symbolic subject where precision and formalism are critical. By expressing ideas in principled operations and numbers, surely the mathematician transcends the vagaries and pitfalls of everyday discourse?

As David Pimm has pointed out (1987: xvii, and see Chapter 2, this volume), such a view disregards two of the essential ingredients of mathematics: people and communication. It ignores the views of many teachers involved in maths education (cf. Brissenden, 1972). It certainly discounts many of the processes and problems of attaining any level of competence in mathematics: mathematics education begins and proceeds in language, it advances and stumbles because of language, and its outcomes are often assessed in language. Such observations could be made of most school curricula, but the interweaving of mathematics and language is particularly intricate and intriguing, and we aim in this book to account for some important aspects of this complex interrelationship.

The concerns, perspectives and methods of the contributors are varied, but they all share an interest in communicative processes and the role of language in mathematics education. In this introductory chapter, I attempt to set the scene for the more detailed discussions which follow by outlining some of the ways in which language and mathematical learning intersect. The issues are presented in order corresponding to the sections of the book. I consider first some aspects of children's early experiences of number-related language – experiences which precede the more deliberated activities of school mathematics. These experiences are of interest to mathematics educators because they influence what knowledge and skills children bring with them to school, and I go on to consider aspects of the relationship between the child's developing knowledge of language in everyday contexts and the language of early mathematics education. I consider some of the specific linguistic demands of the mathematical curriculum, prefacing concerns – ranging from the infant classroom through to secondary education – which are addressed more fully by the contributors. Analysis of linguistic issues in any area of education leads inevitably to the consequences of variations in linguistic experiences and abilities, and I consider these with reference to linguistic or literacy impairments as well as to cross-linguistic differences.

Although the linguistic aspects of mathematics education are regarded throughout this volume as interesting and important, it should be clear that we are not assuming that language is the only matter of concern for the maths educator, nor even that language development and mathematical development are highly correlated. It is worth bearing in mind that most children are very good at learning and using language – they make remarkable achievements in this domain before they commence schooling and in the absence of formal instruction – while very few children take so readily to mathematics. What we are assuming is that much of children's mathematical education takes place in language, and that it is desirable to discover how this occurs and what problems, and benefits, it entails.

This is warranted by the educator's need to talk to (or write for) the pupil in the course of maths teaching (Rudintsky *et al.*, 1981), the pupil's need to *make* as well as receive meanings in mathematics (cf. Murray, 1985), and the sometimes overlooked consideration that the teacher can enter the student's world by *listening* to what he or she says about the journey into mathematics (cf. Mason, 1987:80). For all of these reasons, and others that will be elaborated in the course of the book, the role of language in maths education is crucial. In addition, because maths education makes particular linguistic demands, requiring explanations and questions, precision, extension, reflection and transposition, it has the potential to stimulate and direct aspects of language use and knowledge. As HM Inspectors of Schools (1978) have noted: 'Mathematics lesssons in their turn could make a more substantial contribution to pupil's general language development.'

I turn now to an overview of the order in which some of these considerations have been addressed in this volume. In Chapter 2, David Pimm presents a more general introduction to the notion of 'communicating mathematically'.

Section 2: Language and early experiences of number

Children in Western and many other cultures are exposed to number words and are encouraged to learn them from an early age (Ginsburg, 1977). From the outset, children's first experiences of mathematical tasks of any level are mediated by language (Aiken, 1972; Austin and Howson, 1979; Corran and Walkerdine, 1981; Nesher, 1988; Solomon, 1989; Tizard and Hughes, 1984). However, it is important to bear in mind that language itself is something which children are learning, and which presents enormous challenges in its own right. Hence, in learning about numbers and mathematics through speech, children are faced simultaneously with demanding tasks in different areas. Could these demands impinge upon one another?

There is evidence to indicate that they do. Consider, for example, children's initial exposures to number words and counting. On the one hand, parents have various simplification strategies which they deploy in attempting to promote the linguistic development of their offspring – when addressing young children, they speak more slowly, use relatively basic terminology, with greater emphasis, they point at objects they are labelling, and they provide frequent repetition of key words (Cross and Morris, 1980; Ellis and Wells, 1980; Harris *et al.*, 1983; Messer, 1983; Snow, 1972). On the other hand, there are rules governing the use of number words, such as the order in which they must occur, the one-to-one relationship they have with countables in any particular counting operation, and the fact that the last number in a count tells us how many objects are in the set counted (Fuson, 1988, and Chapter 3, this volume; Gelman and Gallistel, 1978). In a longitudinal study of spontaneous uses of number words by mothers interacting with their young children (from 9 months to 3 years), my colleagues and I found evidence that sometimes they can come into conflict (Durkin *et al.*, 1986b).

For example, consistent with well-established findings in the language acquisition field, we found that mothers' speech in this area emphasized the basics. The number words *one*, *two*, *three* and *four* were the most frequently used, and in declining order. Thus, it appears that children in the early stages of language and numerical development are exposed most often to the most basic elements in the number sequence; indeed, they are exposed to them in simple, rote strings and they are encouraged to practise the strings. However, even in these seemingly simple opportunities to become accustomed with the number sequence, scope for ambiguity and misunderstanding abound and many mismatches occur.

One manifestation of this appears when mothers try to prompt a counting effort by saying the first word ('One') in the hope that the child will produce the next, and so on. In many other contexts of language use, parents say a word in order to elicit a direct *repetition* of that word (presumably reflecting a common lay belief that language is learned by imitation). Now, when the child is exposed to a number word prompt, he or she has no immediate basis for knowing whether a repetition of the exemplar is called for, or a continuation of the string. As a result, many mismatches occur, with the child repeating 'One', or the child responding 'Two' only to be corrected 'No, say "One".' Sometimes, the child misconceives completely what is required:

Mother: Shall we count them? One yellow one.
Child (aged 30 months): Yellow.

(Durkin *et al.*, 1986b:280)

Elsewhere, in order to emphasize the importance of the last word in a count sequence, mothers repeat it, thus: 'One, two, three, four. FOUR.' This may or may not be advantageous to the learner, but note that this common teaching technique, used widely by parents when they are trying to add to their child's vocabulary, happens to contradict basic principles of counting, that the number words are different from each other and that they appear in a stable order: the number word which comes after *four* is not *four*.

Fuller discussions of these and other potential confusions in the early learning of number words can be found in Durkin *et al.* (1986b), and Riem (1985) presents a detailed account of parental teaching strategies in relation to early counting. For the moment, the point is merely to illustrate that even in the early stages of number word acquisition, linguistic factors are important, but their role is not straightforward – a theme that will recur many times throughout this book.

Nevertheless, children do of course begin to count early on (Fuson *et al.*, 1982; Wagner and Walters, 1982) and there has been a great increase in awareness of the importance of the emergence of counting skills in recent years (Gelman and Gallistel, 1978; Fuson, 1988). In Chapter 3 of the present volume, Karen Fuson reviews children's early achievements and illustrates their significance for anyone with interests in the course of mathematical development. She points out that children have to learn to say the number words, that they have to learn to use them to count things, and that they have to understand the cardinal meaning of the final word. Each of these is a complex development and Fuson presents a succinct account of what we know about them as a result of the explosion of research in this area in the last decade or so. She also addresses their integration. As she emphasizes, these counting capabilities are the foundation for the child's subsequent encounters with the mathematical tasks of early schooling.

Alyson Davis (Chapter 4) and Richard Cowan (Chapter 5) consider some of the subsequent encounters. Davis begins by pointing out that our understanding of children's interpretation of mathematical tasks is likely to be of increasing concern in the near future as a result of policy pressures upon mathematics education. In attempting to assess children's mathematical skills, we are faced with a classic problem that we may actually be gauging children's procedures for dealing with adult questioning rather than tapping directly their abilities in the area of concern. This problem has received extensive attention from developmental psychologists in recent years, leading to a recognition that the way in which we pose a question to children can make a lot of difference to the way in which they provide the answer. Davis reviews this work and shows its implications for mathematical educators and testers. She demonstrates that what by adults' standards might appear quite minor changes in the wording of an instruction can affect substantially the way in which young pupils set about a mathematical task – even to the extent of whether they define it as number-related at all.

Richard Cowan dissects the demands of questions involving the seemingly transparent phrase 'the same number'. He shows that a host of linguistic, contextual, conceptual and procedural factors are relevant to an understanding of how children understand such an expression. As with Fuson, he pays particular attention to the role of counting, and argues that learning to exploit counting effectively in *same/different* judgements of number arrays takes several years. He reviews recent experimental work investigating the contributions that sharing and matching tasks might make to facilitating children's progress in this respect. Although the indications are that such interventions can be effective for some pupils, Cowan stresses that children below the age of 7 continue to experience difficulties with these terms. It scarcely needs to be added that this is more than an academic point: it is difficult for any adult to talk for very long about numbers without alluding to their sameness and differences, and Cowan's work reveals that such descriptions cannot be expected to be universally comprehended in the lower primary years.

Another important early task involving the intersection of number and language is learning to understand and use *written* representations of numbers. In Chapter 6, Anne Sinclair examines the beginnings of these achievements. She continues the emphasis upon counting as an integral process in the dynamics of early learning in this area, and considers also the ways in which the everyday environment presents numbers to preschool and kindergarten children. She makes the interesting observation that many encounters with numbers will tend to focus on their nominal or ordinal properties, such as licence plates or house numbers. It is not surprising, she proposes, that children's early explanations of what numbers are and do reflect these incidental but significant experiences. Acquiring a fuller comprehension of place value takes time, and involves an integration of experience, cognitive construction and symbolic knowledge in the course of interaction with others throughout the early years of schooling. Sinclair illuminates the processes and problems underlying these developments.

Section 3: Language and meanings in mathematical education

One of the principal functions of language is to transmit meaning. One of the principal problems of language in mathematics is that the meanings to be conveyed are often complex, and the words we use to convey them are often endowed with other meanings, meanings which may be more familiar to children in everyday language. In Section 3, we consider some aspects of meanings in language commonly used in maths lessons and texts. A large number of new words are introduced to pupils in these and there is evidence that many children experience difficulties with them (Earp and Tanner, 1980; Preston, 1978).

Kevin Durkin and Beatrice Shire (Chapter 7) discuss lexical ambiguity in mathematical descriptions. Our concerns will be familiar to many maths teachers: the vocabulary of mathematics includes a lot of words which have multiple meanings and there is evidence that pupils often fail to interpret the words as teachers intend them. We outline the different types of lexical ambiguity that linguists have classified, and illustrate examples of each type in maths. We list numerous ambiguous words that are used at different levels of mathematical curricula. Drawing on our own work and that of others, we point to evidence that children experience difficulties with this aspect of maths vocabulary when reading, and in direct exchanges or in following instructions.

Further instances of ambiguous terms and phrases are considered in Daphne Kerslake's (Chapter 8) account of the language of fractions and Julie Anghileri's

(Chapter 9) discussion of the language of multiplication and division. In these contexts as well as ambiguity, children have to contend with further examples of the contradictions inherent in some attempts to simplify the task by describing it in concrete terms. For example, children are told early on that 'multiplication makes bigger', a seemingly reliable regulation which proves incorrect and misleading when they reach fractions (and it is interesting to note that even some trainee teachers find it difficult to dispense with this and related misconceptions: cf. Tirosh and Graeber, 1989). Similarly, while it makes intuitive sense to encourage young children to conceive of fractions in terms of familiar objects ('half of a banana', etc.), Kerslake reveals the limits of such a strategy when it comes to dividing a quarter of an apple by two-thirds of a bun. She presents richly informative illustrations of pupils' reasoning about fractions based on transcripts of interviews, which prove an excellent means of soliciting the limits and biases of their understandings.

Like several other authors, Anghileri stresses the importance of children's early, informal experiences of the words used in these tasks, and she outlines everyday contexts which are likely to influence children's understanding. Intuitively, some of these experiences could be expected to be beneficial, though the transition to formal understanding of the operations involved remains demanding. Anghileri goes on to illustrate the complexities of introducing children to multiplication and division in the classroom, with particular attention to linguistic ambiguities and the meaning of symbols.

In Chapter 10, Rita Nolder presents a detailed account of the roles of metaphor in mathematical contexts in the secondary school. Metaphor is one of language's most potent and creative tools, though perhaps one that on first consideration would appear to be the preserve of the literature teacher rather than the mathematician. Nolder establishes clearly that this is an unduly narrow view, and she depicts the pervasiveness and importance of metaphor in mathematics. Her theme is that metaphor is not a superficial decoration or an unwarranted disruption of mathematical discourse, but a basic feature of human communication which offers benefits and problems here as in other manifestations. She shows how metaphor can help and how it can hinder communications, and explores some of the metaphors that pupils themselves construct in mathematics.

Section 4: Word problems

Many mathematical tasks are presented to children as word problems. In a typical example, two or more related pieces of information are presented and the child's task is to supply a missing number by performing the appropriate mathematical operation: 'The book contained 11 chapters, and Mary had read 7. How many more chapters did Mary have to read to complete the book?' Such tasks are important, because they are so widely used in maths education and because they attempt to represent the kinds of problems that call for mathematical skills in everyday life. The demands the tasks make are also very interesting, because they involve an interplay of syntactic, semantic, inferential, temporal and contextual knowledge, and these factors can be found to varying levels of difficulty in different word problems (Kintsch and Greeno, 1985). In Chapter 9, Anghileri notes that even very common everyday words such as *each* and *altogether* can be quite confusing in these contexts. Section 4 presents a detailed examination of the processes of word problem solving by Erik De Corte and Lieven Verschaffel and by Eva Teubal and Pearla Nesher, respectively. Some related issues are also addressed later in a cross-linguistic context by Philip Clarkson (Chapter 20).

De Corte and Verschaffel point out that word problems have been a prominent feature of primary maths education for as long as we are able to trace. They favour an information-processing approach to the characterization of the task demands of word problems and the analysis of children's responses. In the course of a summary of their own extensive research using a wide range of methods, they demonstrate convincingly that word problems involving exactly the same arithmetical operation can actually pose very different levels of difficulty to the pupil as a function of variations in the semantic structure underlying different expressions of the problem. Further, different versions of the problem promote different strategies of solution. The practical relevance of such an approach is at once apparent: a fuller understanding of the types of word problems and the corresponding solutions yields a basis for purposeful interventions to maximize the clarity of the demands of a given task. The authors consider current practices in the presentation of word problems, and maintain throughout the chapter their conviction that progress in research is pointing to a basis for improvements in instruction.

Teubal and Nesher's research leads to similar optimism. They are concerned with the relationship between the order of mention of parts of a word problem and the order of the arithmetical steps the solver has to undertake. We could rearrange the above example in several different orders: 'How many more chapters would Mary have to read to complete the 11 chapter book when she had read 7?'; 'Having read 7 chapters of the 11 chapter book, how many more should Mary have to read to complete it?'; 'How many more chapters would Mary have to read to complete the book, given that she has read 7 and it contains 11?', and so on. If we focus the problem on a different outcome (for instance, if we set the pupil the task of computing the total given the number of chapters read and the number still to be read) we can multiply the possible options. Does this make a difference to the pupil's response strategies? Teubal and Nesher show that it does, and that the kind of solutions obtained vary as a function of the order relationships inherent in the problem and the developmental level of the solver.

Both chapters in this section lend themselves readily to many applications, because word problems are tasks which are set with (ideally) the benefit of planning. Hence, it is possible to control the various stages of difficulty in lesson plans and/or maths texts, and both chapters here present valuable guidelines in this regard.

Section 5: Discussion

Researchers studying language development have become increasingly aware in recent years of the importance of social interaction and conversation as the principal contexts of acquisition (Dore, 1979; Durkin, 1987; Robinson, 1984). Children learn much about language and its functions by means of participation in meaningful interchanges with other language users (Bruner, 1983). In turn, learning *about* communication affords opportunities to learn *from* communication, and another major emphasis of recent developmental psychology has been on the contributions that participation in social activity may make to children's construction of knowledge and understanding (Durkin, 1988). There is now considerable evidence that through social collaboration, children can achieve cognitive insights that they were unable to obtain when working independently. This has been shown in respect of a number of Piagetian conservation-like tasks (Doise and Mugny, 1984) and in mathematical contexts (see Perret-Clermont, 1980; Forman, 1988). It has been argued that educational development itself 'is not simply a matter of individual cognitive growth, but rather a joint enterprise in which shared understandings, terms of reference and forms of discourse are established' (Edwards and Mercer, 1986: 200; see also Bishop, 1985). The evidence is compelling that children can and do learn from social interaction: 'A trouble shared, in mathematical discourse, may be a problem solved' (Wood 1988:210).

This suggests that we need to encourage such interaction (see Brissenden, 1985). To do so effectively, we need to know more about how it may proceed, how to examine it, and how to evaluate its outcomes. We have already touched upon the social interactive processes wherein children begin to make their early discoveries about number words and counting (see also Nesher, 1988; Saxe *et al.*, 1987; Solomon, 1989). In this section, Susan Pirie (Chapter 13) and Celia Hoyles, Rosamund Sutherland and Lulu Healy (Chapter 14) address the role of discussion in mathematical learning by older children.

Pirie presents a fascinating transcript of an attempt by three secondary pupils to solve a written problem concerning the correspondence between relationships among the numbers themselves. She analyses the transcript into a series of progressive episodes, and shows how the language used can provide insight into the mathematical reasoning of the participants. She argues that both what is spoken and what is unspoken in the discourse are critical to 'doing' mathematics, and draws attention to the fact that the kinds of ambiguities and multiple meanings considered in Section 3 can be especially problematic in the context of dynamic interchanges involving several participants, among whom several different interpretations may be held. Despite the thoroughness of her own account, Pirie stresses the multifaceted nature of discourse and encourages the reader to regard her approach not as a rigid procedure but as a flexible framework which can be adapted and amended for application to many other samples of classroom discussion.

Hoyles, Sutherland and Healey draw upon advances in the study of social interaction and learning to investigate pupil interchanges in a vital area of contemporary mathematical activity where discussion might not readily be expected: computer environments. Again, presenting data from rich transcripts of, in this case, dyadic pupil discussions, they show how the analysis of discussion can be used to investigate the ways in which the children negotiate and represent generalizations about the properties of arrays or numerical relations with which they are working. Hoyles *et al.* reveal also that different computer environments (they contrast spreadsheets and Logo activities) influence the nature of the language used and that language serves in different ways to facilitate solutions. As they stress, an important overall conclusion is that computer contexts foster active and task-oriented discussion. Their chapter will be particularly useful to the teacher or researcher who wishes to classify and quantify specific response modes that pupils manifest in generalizing about processes discovered in computer activities.

Section 6: Language, mathematics and disability

So far, we have taken it for granted that mathematics is taught typically to people who have language. But of course there are some pupils who are limited in this capacity, and their mathematical potential and progress are important both in their own terms and as a contrast to normal experiences. In this section, we consider the mathematical experiences of children with three different linguistic handicaps: deafness, dyslexia and language-disorder.

In Chapter 15, Jeffery Barham and Alan Bishop review research into hearing-impaired children's performance in mathematics, and report that the initial prognosis is depressing: deaf children regularly underachieve in this area of their education. They ask why this might be the case and what might be done about it. Among the factors they identify are, once again, ambiguous language – though with some additional difficulties in this context – as well as the dependence of much of maths instruction upon use of logical connectives and other function words that often require complex clauses. However, their conclusion is certainly not that these forbidding obstacles place successful mathematical education beyond the reach of the deaf child and his or her teacher. Instead, they describe an intervention research project which reveals the scope for computer-based mathematical learning as an environment within which deaf children can attain a control and autonomy that often eludes them in more conventional maths work.

As will be seen in Sinclair's chapter, an integral feature of mathematical learning in Western schooling is its representation in written form. In Chapter 16, Michael Thomson considers the plight of the dyslexic child, for whom working in print is a source of difficulty and distress. There is evidence that dyslexic children experience problems with mathematics and lag behind their peers. Although this topic has been relatively neglected in dyslexia research, Thomson draws together evidence which indicates that the problems are not across the board but tend to relate to particular difficulties in perceiving and organizing visual/textual information. He notes too that the kinds of word problems considered in Section 4 can be all the more problematic for these children because they have difficulties in reading the problem in the first place. Drawing on his own experiences in the education of dyslexic pupils, Thomson outlines a number of steps towards improving their performance in mathematics.

Language-impaired children form an heterogeneous population, with each individual presenting a unique set of disabilities and problems. Research into these children's mathematical learning is scant, but it is clear from the approach taken by Chris Donlan and Ella Hutt in Chapter 17 that progress will depend on detailed case studies and the systematic recording of the outcomes of teaching strategies. They present three such studies of three quite different children who face problems with the number system and place value, counting and time concepts, respectively. Their approach is practical: they show how to pinpoint the child's difficulties and they identify operational tactics for guiding and reinforcing learning. It is striking that, like Barham and Bishop and Thomson, their conclusion is that even quite severe linguistic impairment should not be accepted as precluding mathematics education, and they emphasize that an important outcome of efforts in this direction can be enjoyment for both teacher and pupil.

Section 7: Cross-linguistic issues

If linguistic factors are influential in mathematics education, the question arises of whether there are differences in learning and development as a function of the particular language or languages employed. Several important issues stand out here. One of the traditional debates of anthropological linguists, for example, has concerned the controversial Sapir-Whorf hypothesis that 'the language habits of our community predispose certain choices of interpretation' (Sapir, 1921: 162). In line with this hypothesis, it is argued that people think and perceive in ways made possible by the vocabulary and phraseology of their language, and that concepts not encoded in their language will not be accessible to them, or at least will prove very difficult. This hypothesis has been rejected in respect of several areas of concept development (see Dale, 1976: 236ff. for a discussion), and Zepp (1989) reviews studies in which he compared the mathematics performances of pupils and students working in different languages which also provide evidence against the hypothesis. In one of his experiments, Sethoto-speaking children took a maths test in their own language and in English; there was no evidence of superiority in the own-language version of the test at any age level, but older children actually fared *better* in their second language.

Although the Sapir-Whorf hypothesis of constraints upon conceptual opportunities due to the structure of the language does not appear to be well supported, it remains the case that different languages present different challenges to the learner. An important example is in respect of the system governing number words. There is a strong emphasis in several chapters on the role of counting; in

Chapter 18, Karen Fuson and Youngshim Kwon return us to this topic with a cross-linguistic perspective. They compare the ways in which Chinese-based and European-based number words systems are constructed, and examine the consequences for the learner of acquiring either a regular or an irregular system. In a system of the Chinese type, the base ten structure is reflected directly, so that the equivalent of eleven would be 'ten one', twelve would be 'ten two', and so on. In constrast, European languages have various irregularities and inconsistencies in their counting systems beyond ten. The concepts are encoded in European languages, but they are less transparent than in the Asian languages. Fuson and Kwon illustrate many of the differences and then review a growing body of research which indicates that there are repercussions in terms of the speed and fluency with which Asian and Western children learn their number word sequences, and in terms of their progress in elementary arithmetical operations and understanding. They compare also the teaching methods used in Asian and Western schools, and consider how practices in the latter might be improved to ameliorate Western children's linguistic disadvantage.

Hajime Yoshida and Kazuhiro Kuriyama address related issues and continue the emphasis on counting in Chapter 19. They summarize a series of studies conducted in Japan, designed to investigate kindergarteners' representations of the number system, and they consider their results in the context of how numbers and counting are taught in early Japanese education. They advance the proposal that the numbers 1–5 serve as a 'privileged anchor' in early counting and early arithmetical operations, and describe elegant experimental work to support their case. They are not claiming that the privileged anchor is unique to Japanese-speaking children, and it remains for future research to discover how these findings compare with Western replications. Yoshida and Kuriyama show how the anchor is exploited in some aspects of teaching in Japanese schools, and they provide further examples to illustrate Fuson and Kwon's account of the differences between Chinese-based and Western counting sequences.

Another way in which the linguistic circumstances in which pupils study mathematics are important is that they constitute the broader sociopolitical reality surrounding the educational context (Morris, 1978). As a result, the choice of language can be associated with ideological and affective connotations (e.g. see Zepp's, 1989: chapter 12, interesting discussion of the reservations of African students concerning the perceived imposition of colonial culture via Western mathematics taught in European tongues). Whatever one's views on this matter, there are sometimes compelling pragmatic reasons for choosing to teach in a second language. Such circumstances are illustrated in Philip Clarkson's (Chapter 20) account of mathematics in the multilingual society of Papua New Guinea. Here, the sheer numerosity of local village languages makes selection of an official language essential, and the aspirations of a developing nation make English almost inevitable. English, notwithstanding the irregularities of its counting system, has evolved a relatively rich mathematical register, making it a standard tool in this respect (cf. Morris, 1978: and see Pimm, Chapter 2, this volume, for a fuller discussion of register).

Clarkson reviews a number of studies which provide strong evidence of a link between English language ability and mathematical performance in Papua New Guinean pupils. In his own work, summarized here, Clarkson has attempted to locate the particular areas of difficulty, and his tests and detailed error analyses of children's responses to word problems indicate that reading and comprehension problems must be central to any explanation. He discusses the practical implications of this work in terms of experience gleaned with teachers in Papua New Guinea, and provides interesting pointers to the importance and effectiveness of raising metalinguistic awareness in this context.

At several points in the book, we touch on the ways in which children encounter numbers and counting in everyday life. In the final chapter, Judith Kearins describes differences in the ways in which different cultures use numbers. She focuses on two cultures with very different histories and patterns of social organization: Westerners and Aboriginal Australians. She charts the evolution of numbers systems in these respective societies, and shows that their distinct ecologies have led to marked differences in the societies' traditional uses of number and counting. This, of course, is more than a matter of academic interest, as the juxtaposition of the two traditions in contemporary Australia raises many practical issues concerning the adequacy and appropriateness of particular educational systems. Kearins has investigated the number knowledge of Aboriginal and non-Aboriginal Australian children, and she summarizes her findings here. Important differences are clear, and Kearins emphasizes the risk of double disadvantage to Aboriginal children in an educational context which may make unwarranted assumptions about the background knowledge of pupils and thus lay early foundations for a sense of confusion and failure in an alien way of looking at the world.

These, then, are some of the challenges and issues that language in mathematical education presents. Language is critical to many of the processes of learning and instruction, and it confers many benefits in terms of enabling us to articulate, objectify and discuss the problems which the field of mathematics presents. Yet language brings its own rules and demands, which are not always in perfect correspondence with the rules and demands of mathematics; it presents ambiguities and inconsistencies, it can mislead and confuse. Children are developing their linguistic abilities and mathematical competencies, and the relationships between them are not constants but are subject to *intra*- and *inter*-individual differences. It would also be erroneous to suppose that there is some developmental stage or maturation point at which linguistic aspects of mathematical learning cease to be problematic; even adult students manifest difficulties due to the wording or other linguistic properties of mathematical tasks (e.g. Lewis and Mayer, 1987; Peters, 1975; Pimm, 1987; Tirosh and Graeber, 1989).

Hence, language in mathematics education brings together a great many issues. Language itself is the focus of diverse fields of study, all of which are relevant to mathematics. We have touched above on considerations in syntax, semantics, pragmatics, discourse, literacy, sociolinguistics, language impairment, bilingualism/multiculturalism, and others. Despite this variety, we do not pretend to have exhausted the full range of ways in which these two important areas of development intersect, and of course we would not deny that there are many other problems in learning mathematics beyond the linguistic. Nevertheless, the evidence is overwhelming that to communicate mathematically – at any level – you need to communicate linguistically.

Practical implications

As I have stressed, the topic of this volume is of direct relevance to classroom practices. All of our contributors are concerned with the insights that research in their speciality can provide, but also with the implications of that research for aspects of teaching and learning. Consequently, each chapter contains a section in which the author or authors review the practical implications of the work they present. These vary from brief recommendations to quite detailed advice on procedures that the authors have found effective in their own teaching or clinical work. Our goal is not to outline a prescription or dogma, but to offer guidelines towards implementations and explorations that teachers, textbook writers and others involved in maths education may find useful in their own work.

Two general points may be stressed in relation to the practical implications of the material presented in this book. First, the solution is rarely simple. It is certainly likely to be the case that most teaching (and not only in mathematics) is enhanced by the use of clear and effective language; but defining this more precisely proves a major challenge. The fact is that the language of mathematics often *is* demanding and ambiguous – pupils have ultimately to come to terms with this reality rather than to avoid it. Further, attempts to 'simplify' language do not always have the intended consequence: Shuard and Rothery (1984: 132) remark that claims of simple vocabulary in some maths texts translate to 'avoiding the use of words as much as possible!' They also point out that some primary texts written with the express goal of becoming clearer turn out to have higher measured reading ages, with longer sentences and more polysyllabic words. If we could provide a simple procedure for dealing with language in mathematics, we would; in its absence, we offer an abundance of problems and proposals for consideration.

Secondly, the bases for action are many and are promising. As I have already indicated, our contributors are optimistic that practical steps can be taken towards improving pupils' progress even under very handicapped conditions. The thrust of the book is to identify and appraise problems, and then to look for opportunities.

Conclusions

Knowledge is transmitted and advanced via language: face-to-face, in texts and in other media. Children and their teachers write, read and talk about maths. As with most specialist subjects, new terms have to be acquired, old terms used in new ways, and new ways of examining phenomena have to be mastered.

These processes are complex, but their centrality to mathematical education makes them rewarding areas of study. We hope that the following pages will provide some indication of the rewards, and will serve to assist the reader interested in further study, research and practice in the field of language and mathematics education.

Acknowledgements

I am grateful to the contributors for their patient responses to editorial requests, to John Skelton and Pat Lee of Open University Press for support and encouragement of this volume from its conception, to the production staff at OU Press, to Debbie Roodbeen and Sue Davies for secretarial skills and energy, and to Parvin Durkin, for help throughout.

2 Communicating mathematically

David Pimm

Science begins with the world we have to live in. . . . From there, it moves towards the imagination: it becomes a mental construct, a model of a possible way of interpreting experience. The further it goes in this direction, the more it tends to speak the language of mathematics, which is really one of the languages of the imagination, along with literature and music.

Northrop Frye, *The educated imagination*

There are many different relationships that can be highlighted between mathematics and language and my intention in this chapter is to start to explore just a few of them. My primary interest is with the teaching and learning of mathematics, but I feel that it is of crucial importance for all of us to find ways of talking about mathematical activity, which is a very complex phenomenon. One consequence would be to focus attention on particular features of doing mathematics, which could then afford teachers greater insight into what is happening in and between their pupils when in the mathematics classroom.

For me, exploring mathematics in linguistic terms is one such 'seeing' that mathematics education has to offer, though the direct claim that mathematics *is* a language is one that requires careful consideration (aspects of some of the themes mentioned in passing in this chapter are examined in greater detail in Pimm, 1987). Mathematics is not a natural language in the sense that French and Arabic are – for instance, there is no group of people for whom mathematics is their first language. Mathematics is not even a 'dialect' of English (or any other language which can be used for mathematical purposes).

One way of describing the relation between mathematics and a natural language such as English is in terms of the linguistic notion of *register*. Linguist Michael Halliday (1975: 65; my emphasis) specifies this notion as 'a set of meanings that is appropriate to a particular *function* of language, together with the words and structures which express these meanings'. One function to which a language can be put is the expression of mathematical ideas and meanings, and to that end a *mathematical* register will develop.

With regard to English, this development has been taking place since the sixteenth century: Robert Record's mathematical textbooks were among the very first to be printed in English rather than Latin or Greek. An earlier English lawyer and poet, John Rastell (d. 1536), insisted on the potential of English to express complex mathematical ideas and suggested that the privileged status accorded to Latin and Greek in intellectual matters was a cultural rather than a necessary one arising from the attributes of these particular languages.

> The grekes the romans with many other mo
> In their mother tongue wrote workes excellent
> Then if clerkes in this realm would take pain so
> Considering that our tongue is now sufficient
> To expound any hard sentence evident
> They might if they would in our english tongue
> Write works of gravity sometime among
> For divers pregnant wits be in this land
> As well of noble men as of mean estate
> Which nothing but english can understand
> Then if cunning latin books were translate
> In to english well correct and approbate
> All subtle science in english might be learned
> As well as other people in their own tongues did
> (1517, cited in Fauvel, 1987)

Similarly, recent political changes in certain African countries have resulted in a change from one of the colonizers' languages such as English or French being used as the language of education (among other things) to an African language. This exerts among other things a considerable pressure on that language to develop the means for the expression of mathematical meanings. Halliday (1975: 65) also remarks that 'we should not think of a mathematical register as consisting solely of terminology, or of the development of a register as simply a process of adding new words'.

The requirements of expressing mathematical meanings can place strains on a language. One of the most challenging aspects of language study is giving an account of how registers or languages grow. At one level, this can be looking for principles behind the ways new words and expressions are coined (e.g. 'wally' or 'genetic fingerprint'). One important tool in this creative aspect of language is that of metaphor as a means of providing old names for new things – and as such is a conservative force. Important manifestations of this creative aspect of language are concept extension, multiple meaning and metaphor (discussed in more detail in the chapters by Durkin and Shire, and Nolder).

Thus, while providing pupils with opportunities to gain access to the resources implicit in natural language can be seen as a common aim of all teachers (one interpretation of the 'language across the curriculum' idea), a particular aim of teachers of mathematics should be to provide their pupils with some means of making use of the mathematics register for their own purposes. To that end, a mathematics teacher needs knowledge about the language forms and structures that comprise aspects of that register. Part of learning mathematics is gaining control over the mathematics register so as to be able to talk like, and more subtly to mean like, a mathematician.

How does the mathematics register differ from what we might call ordinary English? Instances of such variation include words which have an altered meaning or grammatical function. One elementary example is provided by the number words themselves. In everyday usage, they function like adjectives: one book, two crocodiles, nineteen tower blocks. In mathematical discourse, they become nouns, constituting one of the first sets of mathematical objects, which in turn have properties like being odd or even, or prime. Pupils have to make sense of new language forms like 'two and three make five' and Hughes (1986: 45) has documented some examples of young pupils struggling to make sense of such utterances (Ram, mentioned below, is aged $4\frac{1}{2}$ years).

MH: What is three and one more? How many is three and one more?
Ram: Three and what? One what? Letter? I mean number? [We had earlier been playing a game with magnetic numerals and Ram is presumably referring to them here.]
MH: How many is three and one more?
Ram: One more what?
MH: Just one more, you know?
Ram: (Disgruntled) I *don't* know.

A second example at an older age (11-year-olds, David and Robert) is given by Richard Harvey (RH) (1983: 28).

D: Fifteen's odd and a half's even.
RH: Fifteen's odd and a half's even? Is it?
D: Yes.
RH: Why is a half even?
D: Because, erm, a quarter's odd and a half must be even.
RH: Why is a quarter odd?
D: Because it's only three.
RH: What's only three?
D: A quarter.
RH: A quarter's only three?
D: That's what I did in my division.
RH: Yes, there's three parts in a quarter like on a clock. It goes five, ten, fifteen.

These pupils and adults are clearly engaged in a discussion – but about what? Where are the referents for what they are discussing – to what is the language pointing?

The symbol *is* the object

The symbolic aspect of written mathematics is one of the subject's most apparent and distinctive features. Symbols perform a variety of functions in our culture, and in the context of mathematics can help to show structure, allow routine manipulations to become automatic and make reflection possible, by putting thoughts 'out there' with some stability, compactness and permanence, as objects which may be examined (see Skemp, 1979, for more details). However, the very 'concreteness' of the symbols and the absence of obvious mathematical objects to act as referents, can lead many pupils to believe that the symbols *are* the mathematical objects. The powerful technique of mathematicians of describing algorithms in terms of the attributes of the symbols adds to the potential confusion (e.g. to divide fractions, invert and multiply; to multiply by ten, add a nought; take it over the other side and change the sign). The following two definitions of an even number reflect these two different levels:

A whole number is *even* if it ends in a 2, 4, 6, 8 or a 0.

A whole number is *even* if it can be divided exactly into two equal whole numbers.

Numbers do not have digits, until they are represented in certain numeration systems. In fact, what most primary school children spend most of their mathematical time struggling with are the intricacies of the decimal, place value representation system and not working with numbers and their properties directly.

Walkerdine has worked extensively on the problem of how to understand the role of the teacher in mediating the pupils' introduction and work with mathematical symbols. She has formulated a general and powerful question of particular salience in this area: 'How do children come to read the myriad of arbitrary signifiers – the words, gestures, objects, etc. – with which they are surrounded, such that their arbitrariness is banished and they appear to have that meaning that is conventional?' (1988: 3). This is one place where the presumption that 'mathematics is all around us' can produce difficulties. For us, as mathematically literate adults, the conventional meaning is so firmly 'in' the symbol (and the notion of place value so firmly 'in' the Dienes apparatus, for instance) that surely it is just a matter of looking.

Mathematics, when spoken, emerges in a natural language: when written, it makes varied use of a complex, rule-governed writing system mainly separate from that of the natural language into which it can be read. So, for instance, the mathematical writing system is non-alphabetic and complex clusters of symbols are formed by a wider range of principles than our writing system, where words are formed solely by putting letters next to each other. Such mathematical encoding principles include symbol order, position, relative size and orientation. Use of this writing system, whether control of the formation of the numerals in infant school, or confident manipulation of algebraic forms in the secondary school, have been given a high priority by teachers, and other chapters in this book explore in more detail some of the difficulties of such acquisition (see, especially, Sinclair, this volume; Thomson, this volume).

Practical implications

There are a number of different characteristics and functions of spoken and written language. One use of written language is to externalize thought in a relatively stable and permanent form, so it may be reflected upon by the writer, as well as providing access to it for others. One characteristic of written language is the need for it to be self-contained and able to stand on its own, with all the references internal to the formulation, unlike spoken language which can be employed to communicate successfully when full of 'thises', 'its' and 'over theres', due to other factors in the communicative situation.

One difficulty facing all teachers is how to encourage movement in their pupils from the predominantly informal spoken language with which they are all pretty fluent (see Brown, 1982), to the formal written language that is frequently perceived to be the hallmark of mathematical activity. There seem to me to be two ways that can be tried. The first (and I think far more common) is to encourage pupils to write down their informal utterances and then work on making the written language more self-sufficient (route A in Fig. 2.1), for example by use of brackets and other written devices to convey similar information to that which is conveyed orally by stress or intonation.

For instance, James and Mason (1982: 251) discuss middle-school pupils moving from spoken to alternative written representations of combinations of Cuisenaire rods, focusing directly on questions of re-accessing the original situation from the linguistic representation (in much the same way that Hughes, 1986, has done with much younger children and the 'tins' game) to provide some criteria for judging the adequacy of a given written expression. Repeated access over time is one of the essential functions of written symbols. One example they discuss is how a description of 'pink and white [pause] four times' can be recorded in various mixtures of mathematical symbols and written words.

A second route to greater control over the formal written mathematical language (B in Fig. 2.1) might be to work on the formality and self-sufficiency of the spoken language prior to its being written down. In order for this to be feasible, constraints need to be made on the communicative situation in order to remove those features that allow spoken language to be merely one part of the communication.

Such situations often have some of the attributes of a game, and provided the pupils take on the proposed activity as worthy of engaging with, then those pupils

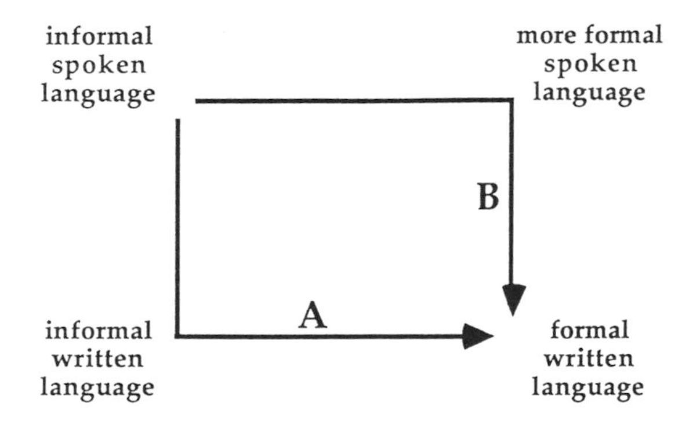

Fig. 2.1. Alternative routes from spoken to written language.

have the possibility of rehearsing more formal spoken language skills. One such scenario is described by Jaworski (1985), where the focus of mathematical attention is a complex geometric poster. Pupils are invited to come out and 'say what they have seen' to the rest of the class, under the constraints of 'no pointing and no touching'. These help to focus the challenge on to the language being used to 'point' at the picture. The situation is an artificial one: in 'real' life, one can often point and this is completely adequate for effective communication. However, if the artificiality is accepted by the pupils, natural learning can take place that would otherwise not have been so readily available. There is an interesting paradox here, one of how quite artificial teaching can give rise to natural learning under certain circumstances.

A second instance of such an approach comes from the contexts of 'investigations', when pupils are invited to report back to the class what they have done and found out. Because of the more formal nature of the language situation (particularly if rehearsal is encouraged), this can lead to more formal, 'public' language and structured reflection on the language to be used. Thus, the demands of the situation alter the requirements of the language to be used. With regard to preparation for coursework write-ups, a prior stage of oral reporting back can help with selection and emphasis.

Reporting back can place some quite sophisticated linguistic demands on the pupils in terms of communicative competence – that is, knowing how to use language to communicate in certain circumstances: here, it includes how to choose what to say, taking into account what you know and what you believe your audience knows. (Some examples of speakers' assumptions of what listeners know can be found in Chapter 13.) A further example of these demands at work can be seen in the study by Balacheff (1988) on 13-year-old pupils' notions of proof, where he asked them in pairs to write down their claims about a mathematical situation to tell another pair what they had found out. By providing them with some plausible justification for them writing a message, he was able to gain access to their proficiency in this matter.

Educational linguist Michael Stubbs writes (1980: 115): 'A general principle in teaching any kind of communicative competence, spoken or written, is that the speaking, listening, writing or reading should have some genuine communicative purpose.' Pupils learning mathematics in school in part are attempting to acquire communicative competence in the mathematics register, and classroom activities can be usefully examined from this perspective in order to see what opportunities they are offering pupils for learning. Teachers cannot make pupils learn – at best, they can provide well thought-out situations which provide opportunities for pupils to engage with mathematical ideas and language.

Conclusion

As Barnes (1976) points out in his book *From communication to curriculum,* communication is not the only function of language. Externalizing thought through spoken or written language can provide greater access to one's own (as well as for others) thoughts, thus aiding the crucial process of reflection, without which learning rarely takes place. In mathematics, language can be used to conjure and control mental images (see, for example, some of the mental geometry activities described in Beeney *et al.*, 1982) as well as provide access to others. Language can be used to point – that is to focus attention – in situations where physical indication is not possible or prohibited for some reason. And despite the mathematician's deliberate (and often powerful) use of 'the symbol is the object' metaphor, it is mental images and acts of attention that are the stuff of mathematics, not marks on paper.

Further reading

Brissenden, T. (1987). *Talking about mathematics: Mathematical discussion in primary classrooms*. Oxford: Blackwell.

Hughes, M. (1986). *Children and number*. Oxford: Blackwell.

Pimm, D. (1987). *Speaking mathematically: Communication in mathematics classrooms*. London: Routledge.

Section 2

Language and early experiences of number

3 Children's early counting: Saying the number-word sequence, counting objects, and understanding cardinality

Karen C. Fuson

Young children in all parts of the world learn to say the number-word sequence used in their own culture. Where body parts (Saxe, 1982) or finger patterns (Zaslavsky, 1973) form the number sequence, children learn to point to these body parts or to make this sequence of finger patterns. Children then use their number-word sequence to count entities; this requires making a correspondence between the number words and the counted entities. Counting is used in three mathematically different situations to determine the quantity in that situation (see Fuson, 1988). The last counted word is used in a *cardinal* situation of discrete entities to tell the *manyness* of the whole set of counted entities, in an *ordinal* situation of discrete entities to tell the *relative position* of one entity with respect to an ordering on all of the discrete entities, and in a *measure* situation of a continuous quantity to tell the *manyness of the units* that cover (fill) the whole quantity. In English and in many other languages, the words from the number-word sequence are used for counting in all of these situations and are also used to tell the manyness in cardinal and measure situations. However, modifications of these number words are used to tell the relative position in ordinal situations; these words are 'first, second, third, fourth, fifth, sixth, etc.'. Thus, at a birthday party, a child might count:

- 'one two three four five' candles on the birthday cake and announce that there are five candles (cardinal);
- 'one two three four five' children in line for a game and announce that 'I am fifth in turn' (ordinal); or
- 'one two three four five' cups of juice poured from a big bottle and announce that 'there are five cups of juice in the juice bottle' (measure).

However, many children entering school do not even know the first five ordinal words (see Beilin, 1975, for a US sample), and primary school children still show difficulties in simple measure situations (Carpenter, 1975; Fuson and Hall, 1983; Miller, 1984).

Although children have much to learn about number during the school years, they are likely to have experienced and used number words in various contexts already, as indicated in Chapter 1. Recognizing these preschool experiences and what children have learned from them is fundamental to understanding how they approach the early stages of formal mathematical education. This chapter presents an overview of the main contexts in which preschoolers use number words. These are *learning the number-word sequence* (i.e. saying the sequence of number words without counting anything), *learning to count* (i.e. saying the number-word sequence in correspondence with entities), and *initial understandings of cardinality* (i.e. using a number word to tell how many entities there are in a set and connecting this cardinal meaning to counting). Preschool children also learn to say number words when seeing the written number symbols (1, 2, 3, ...) (this is discussed by Sinclair, this volume). After summarizing aspects of early progress in these three areas, I will consider how children come to relate these separate meanings and how these meanings are drawn upon by children in their early mathematical work in school. These developments are important and impressive, and I conclude the chapter with recommendations to parents of preschool teachers concerning the kinds of support they can offer.

Learning the number-word sequence

The number-word sequence is originally learned as a rote sequence much as the alphabet is learned. Many of the number words have no meaning initially. The kinds of errors made in learning the sequence seem to depend upon the structure of the sequence of number words (see Fuson and Kwon, this volume). The structure of the English sequence of number words to one hundred is:

(a) a rote list of twelve words;
(b) words 'thirteen' through 'nineteen' repeat the early number words 'three, four, ..., nine' but the irregular 'thir-' and 'fif-' obfuscate this pattern;
(c) a decade pattern of *x*-ty, *x*-ty one, *x*-ty two, ..., *x*-ty-nine in which the *x* words are regular repetitions of the first nine words for 'four' and 'six' through 'nine' but are not regular for two, three or five (i.e. for 'twenty', 'thirty' and 'fifty').

The irregularities in (a) and (b) result in many children learning the words to 20 by rote. Children show understanding of the decade pattern '*x*-ty to *x*-ty-nine' quite early but take a long time to learn the correct order of the decade words (Fuson *et al.*, 1982; Siegler and Robinson, 1982). Most middle-class children between $3\frac{1}{2}$ and $4\frac{1}{2}$ can say the words to ten and are working on the words between ten and twenty. A substantial proportion of middle-class children between $4\frac{1}{2}$ and 6 are still imperfect on the upper teens (the words between fourteen and twenty) but many know those words and are working on the decades from twenty through seventy. Most kindergarten children are learning the decades between twenty and seventy, but a substantial number of them are learning the sequence between one hundred and two hundred.

Children's ability to say the correct sequence of number words is very strongly affected by the opportunity to learn and to practise this sequence. This opportunity may be somewhat less in working-class than in middle-class families (Saxe *et al.*, 1988) and in non-intact middle-class or all lower-class families than in intact middle-class families (Ginsburg and Russell, 1981). Children within a given age group show considerable variability in the length of the correct sequence they can produce.

The incorrect sequences produced by children before they have learned the standard sequence have a characteristic structure. For sequences up to thirty, the characteristic form of most sequences produced by children is a first portion consisting of *an accurate number-word sequence*, followed by *a stable incorrect portion* of from two to five or six words that are produced with some consistency over repeated trials, followed by *a final non-stable incorrect portion* that varies over trials and may consist of many words. The first portion simply consists of the first x words in the conventional sequence of number words and varies with age as outlined above. Almost all of the stable incorrect portions have words in the conventional order but one or more words are omitted (e.g. 'fourteen, sixteen' or 'twelve, fourteen, eighteen, nineteen'). Some examples of typical counting sequences over repeated counts are given in Table 3.1.

Crucially important number-word sequence learning continues long after the child is first able to produce the number words correctly, ranging at least from age 4 to age 7 or 8. New abilities mark qualitative changes in children's mental representation of the number-word sequence. Five levels have been differentiated (see Fuson, 1988, for a summary of these levels and Fuson *et al.*, 1982, for more details).

1. *String level*: the words are a forward-directed connected undifferentiated whole.
2. *Unbreakable list level*: the words are separated but the sequence exists in a forward-directed recitation form and can only be produced by starting at the beginning.
3. *Breakable chain level*: parts of the chain can be produced starting from arbitrary number words rather than always starting at the beginning.
4. *Numerable chain level*: the words are abstracted still further and become units in the numerical sense – sets of sequence words can themselves present a numerical situation and can be counted, matched, added and subtracted.
5. *Bidirectional chain level*: words can be produced easily and flexibly in either direction.

These different levels are marked by increasingly complex sequence abilities: becoming able to start and to stop counting at arbitrary number words, to count up a given number of words, to count backwards starting and stopping at arbitrary number words, and to count down a given number of words. Children also increase their ability across these levels to comprehend and to produce order

Table 3.1 Examples of children's incorrect number-word sequences with accurate portions, stable incorrect portions, and non-stable incorrect portions

Example 1: Age 3 years 6 months
1 . . . 13 19 16 13 19
1 . . . 13 16 19
1 . . . 13 16 14 16 19
1 . . . 13 16 19 16 13 14 19 16 19
1 . . . 13 19 16 14
1 . . . 13 19 16 14 19 16 19 16
1 . . . 13 19 14 16 14
1 . . . 13 19 16 14 19

Example 2: Age 3 years 10 months
1 . . . 12 14 18 19 15 19
1 . . . 12 14 18 19 16 17 18
1 . . . 12 14 18 19 15 17 18 19 17
1 . . . 12 14 18 19 15 16 17 18 19 15 17
1 . . . 12 14 18 19 16 17 12 14 18 19
1 . . . 12 14 18 19 16 17 18 19 17 14 18
1 . . . 12 14 18 19 16 17 18 19 16 17 18 19 16
1 . . . 12 14 18 19 16 17 18 19 16 17 18 19 17 18

Example 3: Age 4 years 2 months
1 . . . 14 16 . . . 19 30 1
1 . . . 14 16 . . . 19 30 40 60
1 . . . 14 16 . . . 19 30 31 35 38 37 39
1 . . . 14 16 . . . 19 30 40 60 800
1 . . . 14 16 . . . 19 40 60 70 80 90 10 11 10 30
1 . . . 14 16 . . . 19 60 30 800
1 . . . 14 16 . . . 19 60 30 800 80 90 30 10 80 60 31 38 39 32 31 34 35 thirty-ten 31
1 . . . 14 16 . . . 19 30 800 60
1 . . . 14 16 . . . 19 30 1 80 90 60 30 90 80 30

Example 4: Age 5 years 2 months
1 . . . 29 60 . . . 69 50 . . . 59 30 . . . 39 90 . . . 99
1 . . . 29 60 . . . 69 50 . . . 59 30 . . . 37

Example 5: Age 5 years 2 months
1 . . . 29 80 . . . 99 90 . . . 97
1 . . . 49 40 . . . 49 40 . . . 49 80 . . . 89 80 . . . 89 90 . . . 97 99 40

Example 6: Age 5 years 9 months
1 . . . 49 30 . . . 39 50 . . . 59 20 . . . 39
1 . . . 39 50 . . . 59 20 . . . 39 50
1 . . . 49 30 . . . 39 50 . . . 59

Example 7: Age 5 years 10 months
1 . . . 59 80 . . . 89 100 . . . 109 70 . . . 79 80 . . . 89 50 . . . 59

Note: . . . means that the intervening words were said correctly.

relations on the words in the sequence (see Cowan, this volume; Fuson, 1988). Children's construction of these developmental representational levels depends less on particular features of the sequence than does learning the correct number words, so these levels seem to appear fairly widely: in children in the Soviet Union speaking Russian (Davydov and Andronov, 1981), in Israeli children speaking Hebrew (Nesher, pers. comm. 1980), in New Guinea Oksapmin children with a body number-word sequence (Saxe, 1982), and in US children speaking English (Fuson *et al.*, 1982). However, instructional practices may affect the manifestation of these levels (e.g. Hatano, 1982).

Learning to count

In order to count objects distributed in space, one must match each object with a number word. To do this, some sort of indicating act (some variation of pointing or of moving objects) that has both a temporal and a spatial location is needed. The indicating act serves as a mediator and simultaneously creates two correspondences: the correspondence in time between a spoken number word and the indicating act, and the correspondence in space between the indicating act and an object. The required matching between the spoken word and the object in space – the word–object correspondence – then derives from the separate temporal and spatial correspondences involving the indicating act. For counting to be correct, the *local correspondences* associated with each indicating act must be one-to-one: one word must correspond to one indicating act and one indicating act must correspond to one object. The *global indicating act–object correspondences* over the whole counted set must also be one-to-one: every object must be indicated and no object can be indicated more than once. This requires some method of remembering which objects have been counted. Finally, for the counting to be correct, the words said must be the standard number-word sequence.

The kinds of correspondence errors 3- to 5-year-old children actually make in counting and the variables that affect these errors are described in Fuson (1988, chs. 3–6; see also Gelman and Gallistel, 1978). Of the 13 types of errors made by children when counting objects in rows, only 5 are made very frequently. Two of these violate the *word–point* correspondence: in *point–no word* errors, children point at an object without saying a word and, in *multiple words–one point* errors, they say two or more words while pointing once at an object. Two other common errors violate the *point–object* correspondence: in *object skipped* errors, an object is skipped over without being counted and, in *multiple count* errors, an object is counted and then immediately counted again (it receives a word and a point and then another word and a point). The fifth common error violates *both* correspondences: in *multiple points–one word* errors, an object is pointed at two or more times while one word is said.

Children aged 3–3½ make a considerable number of these common kinds of errors. The rate of most errors drops at least somewhat at age 3½ and continues to drop at every half year after that. Five-year-olds have quite low rates of local correspondence errors when counting objects in rows but do still make errors that violate the global indicating act–object correspondence when counting complex non-linear arrays (they recount some objects and fail to count some objects). The percentage of children making at least one error of a given type generally drops more slowly than does the rate of making that error, indicating that learning to count primarily involves gradual improvement rather than sudden insights in which a given error type drops entirely out of a child's repertoire.

Children sometimes move their finger along a row of objects saying words without really pointing at objects (a *skim* error) or produce a flurry of words and of points directed generally but not specifically at the objects (a *flurry* error). Both of these types of errors, as well as *skipped objects* errors, are particularly affected by the effort with which children count. When children are asked 'to try really hard', these errors decrease considerably.

Both parts of counting – saying the words and pointing – begin to undergo internalization at the end of the preschool years. This internalization continues into the primary grades and may be accelerated by a teacher who actively discourages children from counting to find addition or subtraction answers. Children begin to count silently, though lip movements are still often evident, and they point from a distance and finally may use eye fixation instead of pointing. Especially initially, children may make more counting errors when counting is internalized. Because counting is crucial for children's understanding of addition and subtraction of small numbers (see Fuson, in press, for a review), it is important for teachers to allow and even encourage children to count.

Counting is a very complex activity. However, even by the age of 3, counting is very organized and usually exhibits the general structure of mature effective counting: it has a recognizable structure of word–point and point–object correspondences, and most of the frequent errors violate only one of these correspondences. Young children can count objects in rows quite accurately: children aged 3–3½, 3½–4 and 4–4½ had correct correspondences on 84%, 94% and 97% of the objects in rows of 4–14 objects (Fuson, 1988). Accurate correspondence does fall considerably for large disorganized arrays, because most young children do not use adequate strategies for remembering which objects they have already counted and so global as well as local correspondences are violated. Gelman and Gallistel's (1978) characterization of preschool children's early counting as governed from the beginning by the One–One Principle (every item in a set must be assigned a unique tag) is a simplification of the nature of counting and of children's counting correspondence behaviour (there are many different aspects of correspondence to understand and to carry out), but their general position that counting is structured and organized and displays considerable competence is supported.

Initial understandings of cardinality

Children initially count without any cardinal outcome for the counting. If asked 'How many candles?' just after counting five birthday candles, many 3-year-olds and some 4-year-olds will count again (they may recount many times to repeated how-many questions), or give a number word other than the last word said in counting, or say several number words. These children do use number words with cardinal meaning in non-counting situations. They subitize (immediately recognize and label) very small numbers of one, two, and possibly three or four objects, and they may imitate and independently repeat cardinal uses such as 'I have five people in my family'. But they cannot yet connect counting to these cardinal meanings: they cannot make a count-to-cardinal transition in word meaning from a counting meaning of the last counted word as indicating the last counted object to a cardinal meaning of that word as telling how many objects there are in all. Most 2- and 3-year-olds do remember the last counted word, and therefore the failure is a conceptual and not just a memory problem.

Children make this fundamental connection in different ways (see Fuson, 1988, for a review of different positions on how children come to make this connection and for data concerning the validity of these different positions). Initially for many children, this connection may reflect only a last-word rule that the last counted word is the answer to a how-many question; that word may not have a cardinality meaning referring to the whole set or to the cardinality of the whole set. Once a child answers a how-many question with the last counted word, that child will continue to do so fairly consistently across trials and across different set sizes. Some children may count a small set, also subitize that set, and see that the subitized word and the last counted word are the same (Schaeffer *et al.*, 1974; Klahr and Wallace, 1976). It may take only one such example for children to generalize this pattern to a last-word rule and use it on non-subitizable larger sets. Children may also make this last-word connection if they experience some event that makes the last counted word particularly salient (makes them notice the last counted word) or if they are told that the last counted word tells how many. Finally, preschool children's recency bias (their tendency to answer a multiple-choice question with the last provided choice) or an auditory 'echoing' of the last word may prompt the child to hazard the last counted word as a guess to the how-many question.

Many 4- and 5-year-old children do make count-to-cardinal transitions after counting a set of objects: when they say 'One, two, three, four, five. There are five candles,' the five refers to all the candles and tells how many candles there are. However, this cardinal meaning is still not mature. Children may still be misled by appearances in equivalence situations and say such things as 'This five candles (a longer row) is more than this five candles') for discussions of equivalence and order situations, see Cowan, this volume; Bryant, 1974; Fuson, 1988; Piaget, 1952). They also may not be able to solve certain kinds of addition and subtraction situations (see Baroody and Ginsburg, 1986; Briars and Larkin, 1984; Carpenter and Moser, 1984; Fuson, 1988, in press; Riley *et al.*, 1983; Steffe *et al.*, 1983; Steffe and Cobb, 1988). Children's understanding of cardinality continues to grow throughout the primary grades (see Fuson, 1988, in press, for reviews of this increasing understanding; Sinclair, this volume, discusses the implications of cardinality for learning to write numbers).

Early relationships among sequence, counting and cardinality knowledge

Gelman and Gallistel (1978) identified three principles that direct young children's counting: the stable-order principle ('the tags used must be drawn from a stably order list'), the one–one principle ('every item in a set must be assigned a unique tag') and the cardinality principle ('the last tag used in a count represents the cardinal value of the set') (definitions are from a recent statement of the principles by Gelman and Meck, 1986). They reported that the developmental relationship among these principles was: stable-order before one–one before cardinality, i.e. children learn a stable list of number words before they can make correct word–point–object correspondences before they can give the last counted word as the answer to a how-many question. Research concerning the relationships among these three aspects of counting is reported and reviewed in Fuson (1988). No support was found for the Gelman and Gallistel relationship except for very small sets. Instead, this relationship was found to depend very heavily on the size of the set counted:

- for sets of 2, 3, and 4: stable-order and one–one before cardinality;
- for sets of 4 through 7: stable-order and one–one before cardinality *or* stable-order and cardinality before one–one;
- for sets between 7 and 16: stable-order before cardinality before one–one; and
- for sets above 16: cardinality before stable-order before one–one.

These relationships are sensible because it is easier to learn a stable list of words or to make correct word–point–object correspondences for a small set than for a very large set but the cardinality principle is a rule that can easily be applied across different set sizes and does not have to reflect understanding of cardinality (and thus can be independent of a correct one–to–one correspondence between the words and the objects). The Gelman and Gallistel result is widely discussed and accepted but the order of their principles seems to be the result of limitations in their testing procedure (Frye *et al.*, 1989).

The integration of sequence, counting and cardinal meanings

Children enter school with strong counting capabilities. The sequence, counting and cardinal meanings of number words form their conceptions of numbers. They use these meanings to make sense of numerical situations in school maths. They approach addition and subtraction situations by counting out objects for the first number, counting out objects for the second number and adding them to or taking them from the first objects, and then counting the total or the remaining objects. These counting abilities enable children to solve a wide range of addition and subtraction situations (see the discussion by De Corte and Verschaffel, this volume), although as Davis shows (this volume), children still have much to learn about the contextual demands of such operations.

The different qualitative levels in children's mental representation of the number-word sequence discussed above indicate important changes in children's understanding of addition and subtraction situations. These changes manifest the child's increasing integration of counting, sequence and cardinal meanings of number words (Fuson, 1988; Steffe *et al.*, 1983; Steffe and Cobb, 1988). At the *unbreakable list* level possessed by most children entering kindergarten, children are able to use counting to determine how many entities there are in a given situation and are able to add by counting all the entities representing the addends and subtract by separating some objects from all the objects. At the *breakable chain* level, children are able to count on with entities, starting the sum count of entities by beginning with one of the addends (e.g. for 8 + 5, counting five more words from eight – '8, 9, 10, 11, 12, 13' – instead of beginning with one), and they can subtract by counting up to (e.g. for 13 – 8, counting '8, 9, 10, 11, 12, 13' to find the answer 5 words counted up from 8 to 13), by counting down from (e.g. for 13 – 8, counting down 8 words from 13 to end at the answer 5), and by counting down to (e.g. 13 – 8, counting down from 13 to 8 to find the answer 5 words counted down). At the *numerable chain* level, the number words can represent the addends and the sum for addition and subtraction situations, and children thus can solve addition and subtraction problems by using the efficient sequence procedures counting on, counting down from, counting up to, or counting down to without needing precounted sets of entities to keep track of the counting; they keep track of the words counted up or down by extending fingers, or using auditory or visual patterns, or by double counting ('nine is one, ten is two, . . ., thirteen is five'). At this level sequence, counting, and cardinal meanings become integrated within a single mental representation of the number-word sequence, as they are for readers of this chapter.

Practical implications

It is important for teachers and parents to realize how very separate and contextually bound all of the different number-word meanings are initially for young children, what an impressive intellectual feat children achieve in constructing and then relating these separate number-word meanings, and how very crucial counting is to children's understanding of number. Children can best be prepared for school maths by having preschool teachers, parents and caretakers who help children take joy and pride in their noticing and labelling numerical aspects of their environment and in their learning of the number-word sequence and who facilitate children's enthusiastic counting of all sorts of things in many different situations. Children can move from counting single sets to counting two sets in order to add them. They can match the objects in two sets or count them to find out which has more and how many more it has. These numerical activities can arise in children's everyday activities ('We have four plates on the table. How many cups do we need if each plate gets a cup?'), in games ('I get to move six spaces!'), and in classroom or family routines ('Let's say the numbers out loud together while we wait for this traffic light' or 'while we line up to go outside'). The focus should be on facilitating children's seeing and talking about many different numerical situations and not on formal number 'lessons' or on pages and pages of workbook problems. Some examples from diaries I wrote to my own two daughters about their early years are given in Table 3.2 to provide a flavour of the range of situations in which numerical conversations can arise (more examples are given in Fuson, 1988, and also see Durkin *et al.*, 1986b; Saxe *et al.*, 1988).

Because so much important numerical learning occurs after children can count correctly, numerical activities are as crucial after children have learned to say the correct sequence and to make accurate counting correspondences as before. The number-word sequence and the counting of objects needs to become very overlearned in order to move through the developmental levels of counting, cardinal and sequence relationships. Many counting activities for many different addition and subtraction situations, and reflection about, and discussion of, alternative solution procedures for these situations can help children move through these developmental levels. Children need to be exposed to many different meanings of addition and subtraction, not just addition as getting more and subtraction as taking away (see De Corte and Verschaffel, this volume, for these alternative meanings). Because counting forward is much easier than counting backward for children, subtraction needs to be given a counting up 'how many more?' meaning ('One meaning of $6-4$ is how many more do I have to add to 4 to get 6?') and not just a take-away meaning. Children move from the usual take-away meaning to counting down for subtraction and make many more mistakes in counting down than in counting up. Finally, many interpreters of Piaget's work on number conclude that children must understand conservation of number (see Cowan, this volume) before they can understand addition and subtraction and thus suggest delaying work on addition and subtraction until after children understand conservation of number. However, understanding conservation of number is a relatively late development that follows the *numerical chain* sequence level. Thus, much sequence, counting and cardinal understanding, and solution of many different kinds of numerical problem situations, can precede conservation of number.

Conclusion

Young children initially learn number words in several mathematically and functionally different situations and thus the same number word comes to have different meanings. The major number-word situations during the preschool and

Table 3.2 Examples of young children's uses of number words in the home

Age	*Diary entry*
1:11	Tonight we counted steps going up to bed (we usually do). We were on nine and you said, 'One two fee four five six'. First time so many were correct.
2:0	'One two three eight jump'. (You are into counting and jumping off the hassock or even the couch.)
2:6	Two tomatoes were on the table. You said and acted out the following: 'One tomato from two tomatoes leaves one. Two tomatoes from two tomatoes leaves no.' I asked you what no tomatoes from two tomatoes was: 'Two tomatoes.' 'Sesame Street' does things similar to the first two sentences.
2:7	Putting prunes back into a box, you correctly counted them up to nine. When asked how many prunes: 'three' (your standard 'how many' answer at this point: three eyes, etc.)
2:10	I cut your peanut butter sandwich in half and then into half again. You watched and said, 'Two and two make four.'
	You just asked for four olives (you love olives!). Your father gave them to you, and you said, 'Two and two make four.'
2:11	I want manier than five.
3:5	You were typing and saying as you typed the numbers (they are in order on the typewriter): 'one, two, three, four, five, six, seven, eight, nine [pause]. I need a ten.'
3:10	'What's four and four?' I said to count on your fingers, 'One, two, three, four. One, two, three, four. One, two, three, four, five, six, seven, eight. Four and four make eight.'
4:2	Walking to my office with you, I asked you how many chairs were in my office. You said, 'Four'. (This was the use of mental imagery; you said you counted the chairs in your head. There were four chairs in my office, one at my desk and three at a table.)
4.9	When we were preparing bags of popcorn and peanuts for your (early) school birthday, you counted kernels of coloured popcorn. In great excitement and wonder: 'Ooh! I counted up to one hundred and two!' You refused to make piles of ten. You finally quit at 150.
	You played this game with your sister (then 2:4): You put coloured beads or other objects out on a cloth and then asked her questions. 'Erica, what does it make adding one and one?' Erica (without looking at the beads): 'Two'. You would put out the correct number of objects. Erica answered somewhat randomly, rarely looking at the objects.
5:7	Conversation with you in the bathtub: You: 'How much is seven and seven?' I held up seven fingers and had you hold up seven fingers. You counted. You still usually will not count on from the first number; you need to count the fingers for the first number. You asked and we did nine and nine, four and four, six and six, five and five, and then ten and ten (you did that with your fingers and toes). You said twenty. I said it was called that because it was twin tens – two tens. You said, 'I know' and thought for a bit. Then you said, 'There's a zero to make the ten and the 2 to make the [pause] two tens.' You asked what twenty and twenty were. We used all our fingers and toes. You

Table 3.2 (*Continued*)

Age	*Diary entry*
	counted on from twenty by ones. Got forty. I asked you how many tens were in forty (pointing to our fingers and toes) – it took a bit of focusing for you to see the tens. You said, 'Oh, the zero for ten and the 4 for four tens.' You drew a 2 in the air and asked, 'Is it this way or the other way [backward]?' You still make many of your numbers backward – but not letters. I asked you how you knew where to move your arm to draw a 2. 'I don't know.' 'Do you see a picture of the 2?' 'Yes, I see an invisible 2, and I just draw it.' You brought home a Montessori numeral sheet today. It was divided into 1-inch squares and was ten squares by eight squares. The teacher had written the symbols in order from 1 through 10 across the top row. You then wrote seven more rows of numerals under them. Your 1, 3, 4, 5 and 7s were mostly quite good. You were struggling with 2s; they turned more and more into Zs as they went down the page. You had trouble with the loop in the 6. The reverse direction cross-overs in the 8 were really difficult for you, so you tried different strategies on the 8s. You tried partial cross-overs and overlapping loops and then settled for a circular top and a U fastened to the bottom of the loop. Your 10s were too scrunched because you started the zero too close to the 1; you went to the right and then when you looped back to the left you ran into the 1. You also started the zero at the bottom. But overall quite a good job; everything was recognizable.
6:0	You know most of the double sums (3+3, 7+7, etc.) right away. I asked you some tonight and then asked nine plus nine. You closed your eyes and scrunched up your face and thought and then said, 'Eighteen.' I asked how you had figured it out. 'Well, I knew eight plus eight was sixteen, and I knew there had to be one in the middle, so it was eighteen' (i.e. you knew the double sums went up skipping every other number). I then asked you five plus seven. You thought for a while and then said, 'Now don't ask me to describe it because it is very difficult.' You closed eyes, etc., and after a while said, 'Twelve'. I was surprised. 'Could you give me a clue?' 'Well, I had one five and there was another five in seven with two left over, so that made one was eleven and two was twelve.' Big smile. Me too.

Note: The age is given in years and months, so 1:11 is 1 year 11 months old.

primary school years are sequence, counting and cardinal situations. Children learn a great deal about each of these kinds of situations and gradually integrate sequence, counting and cardinal meanings within a single powerful and flexible mental representation of the number-word sequence. Use of this integrated number-word sequence permits them to understand and solve addition, subtraction, multiplication and division situations involving numbers less than one hundred.

Further reading

Fuson, K.C. (1988). *Children's counting and concepts of number*. New York: Springer-Verlag

Fuson, K.C. (in press). Research on learning and teaching addition and subtraction of whole numbers. In G. Leinhardt and R. Putnam (eds) *Cognitive research: Mathematics learning and instruction*.

4 The language of testing

Alyson Davis

Internationally, the ways in which we teach and assess mathematics are becoming topics of urgent public debate and policy review. In the UK, for example, the publication of the Cockcroft Report in 1982 put mathematics at the forefront of the curriculum and the implementation of the National Curriculum in England and Wales is likely to reinforce this further. The 1990s should see further change for teachers as they become involved not only in spending a good deal of time in teaching mathematics but also in the assessment of the mathematical skills of their pupils. Given this surge of emphasis on both the teaching and assessment of mathematics, it is an appropriate time to take stock of what we know about children's mathematical skills and how to assess these.

The focus in this chapter is the role of language in testing children's understanding of maths in the early years of schooling, drawing on some recent research in developmental psychology. The research comes from two areas: first, the role of language in testing and, secondly, work on young children's mathematical understanding. The conclusions from these two areas are complementary and together form the beginnings of a story which has direct links to how we might best proceed in both teaching mathematics to young children and monitoring their learning.

Developmental psychology, teachers and testing

The relationship between developmental psychologists and teachers is notoriously ambivalent. This is hardly unexpected given that the aims of these two groups are often quite separable. However, within the area of testing children's knowledge, teachers and developmental psychologists have potentially very similar concerns. Testing and assessment have different meanings to different people depending on their aims and objectives, but at the most general level, testing and assessment refer to methods of finding out what children know and using this information for future planning and decision making.

There is a useful distinction to be made between norm-referenced testing (where children's scores on a standardized test are used to allow comparisons between children) and criterion-referenced testing where children are tested against some preset criterion (e.g. being able to count to ten). With criterion referencing then, the comparisons are less about looking at child against child and more concerned with finding out what individual children know. In the past, a good deal of formal educational testing has been norm-referenced, since most school examinations have been of this kind, but the new National Curriculum is committed to take a more criterion-referenced approach. As a result, a common ground is established between teachers and developmental psychologists.

Research on young children's thinking has more in common with criterion-referenced testing than norm-referenced testing, and therefore the day-to-day assessments that teachers make of children will resemble far more the kinds of tasks that psychologists have used to assess children's understanding.

Testing as an interactional setting

In many ways, one might expect that the testing of mathematical skills should be less problematic than assessing other skills such as reading or language. Mathematics is precise and context free and, as such, making judgements about whether a child's answer is right or wrong is in principle easier. Agreement on what constitutes a 'correct answer' is typically less contentious than in other disciplines.

The problem of assessment, however, is not simply one of whether or not a child knows the 'right' answer, but in knowing the basis on which a child arrives at any answer regardless of its appropriateness in the eyes of an adult. For both teachers and developmental psychologists, understanding why children answer our questions and tests in the way they do is a fundamental issue. As such, testing children's mathematical knowledge is open to much the same pitfalls and benefits as testing other forms of knowledge.

Questions and answers

Testing children's knowledge invariably involves us in looking at children's language. For even if we do not expect children to give a verbal reply in a test situation, most tasks which we present to children involve the adult giving a series of instructions and asking questions in some form or another. Questions and answers are nothing new to young children, they are part of their everyday lives both at home and school, although the different functions of questions in different contexts are now well established (e.g. Tizard and Hughes, 1984).

Typically, what may concern a teacher is how to interpret wrong answers or children who remain silent in response to a question. Ironically, from a psychological point of view, it turns out that the problem is not so much that children may give no answer but that they answer our questions at all! Hughes and Grieve (1978) provide an instructive illustration. They asked children aged between 5 and 8 years, some non-sensical and (from an adult's point of view, unanswerable) questions. For example, 'Which is heavier, red or yellow?', 'Which is bigger, milk or water?.' The children answered these bizarre questions and, moreover, they backed up their answers with justifications such as 'milk is bigger because it comes from cows'. While the older children showed some signs of amusement, indicating they they appreciated the ridiculousness of these questions, the younger children were confident in their answers.

Findings like these serve as poignant reminders that the question and answer situation does not provide us with privileged access to children's understanding. Had the question been of the form 'Which is darker, red or yellow', we might have been tempted to infer that the children (depending on their answers) had a greater or lesser understanding of hue. However, given that the questions posed by Hughes and Grieve were unanswerable, the children's answers reveal more about how children interpret adult language than they do about any particular underlying knowledge which the question is supposedly testing. Stated simply, children do not necessarily interpret our questions in the way that we had intended.

This point becomes crucial when we consider situations in which we are testing children's knowledge or skills. It suggests that before we can use children's answers to infer knowledge or ignorance, we have to take every precaution to ensure that the child's interpretation of the question is as we had intended.

Donaldson (1978, 1982) points out that young children not only know less about language than adults and older children, but that they also rely more heavily on context to derive meaning and lack the experience to know when it is and when it is not appropriate to take adults' language literally and ignore the context.

Fortunately, there is now enough research evidence available on which to outline some of the key factors influencing children's interpretation of language in testing contexts, which makes our task in assessing knowledge a more manageable one. One area of research which has provided a wealth of useful data is children's performance on Piagetian tests of conservation. Conservation or the understanding of invariance is of particular interest here for two reasons. First, Piaget saw conservation of number as being at the heart of children's mathematical understanding and, secondly, the conservation task has taught us a good deal about the role of language in a testing situation.

Cowan, in the following chapter, discusses the conservation task and its relationship to Piaget's theory of number development in more detail. For present purposes, I shall simply remind the reader of the basic procedures. In a standard number conservation task, the child is shown two rows of objects which are equal in number and which are in one-to-one correspondence. The child is asked whether there are the same number in each row and the child agrees. One row is then spread out by the adult so that it appears longer but without the actual number being altered. The question about equality of number in the two rows is then repeated. Young children below the age of 6 or 7 typically claim that there are now more objects in the row which has been spread out than in the unaltered row.

This failure to conserve number was, for Piaget (1952), a fundamental symptom of young children's illogical thinking. However, in recent years, attention has been drawn to the importance of children's interpretation of the language in conservation tasks as explanations of their apparent failure in such tasks. Here, then, we have a task designed to tap a fundamental mathematical concept being scrutinized for its linguistic components.

The conservation task as social interaction

The conservation task is an interactional setting. It involves an adult interacting with children and asking them questions in relation to the adult's actions of setting up rows of objects and moving them about. The traditional task involves not only asking children a question, but also repeating that question – once before the transformation of one row and once following the transformation. The question is the same in each case, asking the children to assess the numerical similarity of the two rows of objects. But many have argued (Donaldson, 1978, 1982; Samuel and Bryant, 1984) that there is every reason to suppose that children change their interpretation of the conservation question despite the wording of the question remaining the same.

Donaldson has proposed that children do not just interpret the words in isolation, but that they interpret the adult's intentions in asking the question and in doing so assume the transformation (which is in fact irrelevant, since the number is not altered) to be the very topic of conversation. Because the adult repeats the test question following the alteration of the length of the row, the child supposes that he or she should answer in terms of that change. Indeed, it is now well established (Donaldson, 1982; Rose and Blank, 1974; Samuel and Bryant, 1984) that by avoiding the repeated question or by making the transformation appear accidental (McGarrigle and Donaldson, 1975) or incidental (Light *et al.*, 1979; Light 1986), children who fail to conserve number on a standard Piagetian task will succeed in significant numbers on these modified tasks. Other research has shown that children's own explanations of what they believe to be the purpose of the adult's questions are closely related to whether they give conservation responses or not (Pratt, 1988). Extensive evidence, then, points to the importance of considering not only the language used in this type of task, but the way it is used and its effects on children's interpretations of the task as a whole.

So far, we have considered the role of language in relation to testing children's general understanding of tasks presented in a one-to-one, adult–child testing situation. Children's dependence on contextual information has shown that there is no direct means of assessing information; rather, children's understanding of the test question and their interpretation of the task as a whole all add up to a more complicated picture of assessing knowledge. In the next section, I turn to some recent research which looks directly at children's mathematical thinking.

Context and the development of mathematical understanding

Piaget provided an underlying theory of the development of mathematical thinking which made its way into teacher education and curriculum practice. But Piaget's message was double-edged. On the one hand, his work became a catalogue of the weaknesses of children below 7 or 8 years of age in terms of the cognitive constraints on their mathematical ability. On the other hand, he saw mathematical and cognitive development as being closely linked. Yet children find mathematics difficult and we need a satisfactory account of why this is the case. Much of the research in developmental psychology in recent years has shown that where Piaget mis-stated the children's case, it was in the direction of underestimating their skills and abilities rather than overestimating them. This, then, has to be reconciled with the contention that mathematics is not easy, nor does it come naturally to all but a few. The paradox can be resolved in part by considering recent findings on children's mathematical skills in relation to the pioneering work on children's contextual sensitivity.

Research by Martin Hughes (1981, 1983, 1986) has explored children's mathematical skills in relation to Piaget's tests of mathematical thinking and more recent accounts of children's performance on these tasks. We know that even young children may show evidence of number knowledge, such as conservation if the task is 'embedded' (Donaldson, 1978) in a meaningful context which allows the child to understand the adult's intentions and purpose of the task. Children show surprising skill when this context-dependent 'embedded' thinking is called for. However, mathematics itself requires a very different type of thinking – it is precise and context-free, and to be able to deal with mathematics the child's task is to take on board a very different type of thinking which Donaldson terms 'disembedded'. Disembedded thinking requires that children must forego their typical reliance on context on which their normal thinking depends and instead deal with a system that by definition is context-free. It is this dilemma which Hughes and others have addressed empirically.

In the research I shall discuss, the focus is on those aspects in which the language of testing is most implicated. Take the following example of a 4-year-old being questioned by an adult (Hughes, 1983: 211):

Adult: How many is two lollipops and one more?
Child: Three.
Adult: How many is two elephants and one more?
Child: Three.
Adult: How many is two giraffes and one more?
Child: Three.
Adult: So, how many is two and one more?
Child: [looks adult straight in the eye] . . . Six.

In this situation, the child appears to have no difficulty with simple addition as long as the problem is stated within a specific topic – lollipops, elephants and giraffes. As we might expect on the basis of the developments reviewed by Fuson in the previous chapter, he can count and can use his knowledge of the counting sequence effectively. However, when the same question is asked without a referent, the child stumbles and gives a confident but incorrect answer. Can it really be the case that the use of a single word appears to make the difference between arithmetic competence and arithmetic ignorance? If it were, then the implications for the way in which we test number knowledge would be enormous. There is also another issue at stake, i.e. in order to deal with formal maths as it is taught and assessed in school, children will have to deal with arithmetic problems which are not embedded in context, and therefore they must at some point realize that arithmetic can be performed in the absence of any referents.

This problem has been tested experimentally by Hughes (1981). Preschoolers were given simple arithmetic problems (such as how many is one and two?) which differed in terms of the amount of context given to the child. All of the problems were presented verbally, but in one case the children were shown real objects (bricks being added to boxes) and in others they were given 'hypothetical situations'. For example, 'If there were three children in a sweet shop and one more child came in, how many children would there be altogether?' And, finally, children were asked the questions in formal code: 'How many does three and one make?'

The results were revealing. As one might expect, the children did slightly better when the question was posed with real objects than in the other two instances. However, the surprising finding was that the most significant difference between the results was comparing performance on the formal code and other tasks. Whereas less than 10% of the formal code questions were answered correctly, over 50% of the children could answer the hypothetical situation questions correctly. Thus, it is not context in the sense of real objects that the children can see or touch which they need to be able to solve the problem, since the children can answer the hypothetical problems which are purely verbal.

Davis and Lo (1986) have also gathered data which support the finding that posing arithmetic problems using a single word or imagined situation greatly improves the children's performance compared with the formal code context-free question. The finding is a very important one because it points to the source of the children's difficulties being something other than cognitive ability. The fact that the children can answer the hypothetical question shows that they know how to do the arithmetic but that they need to rely on context to operationalize their skill. This stands in marked contrast to Piaget's ideas in which the children's problems were accounted for by some conceptual deficit. Instead, a new distinction can be made, whereby children's knowledge and their understanding of when and how that knowledge should be deployed are two separate problems (Bryant, 1985; Davis *et al.*, 1985).

If we consider this evidence in terms of assessing and testing children's mathematical skills, it might be taken in two quite different ways. One argument would be to suggest that given the importance of disembedded thinking in mathematics, children can only really be considered to understand arithmetic if they can answer problems set in formal code. This assumes that for most mathematical purposes, being limited to answering problems only in certain contexts is mathematically worthless.

The problem with this view is two-fold. First, it discredits the knowledge the children have when they can answer embedded questions and, secondly, in doing so it tells us little about how to help the children to learn how to deploy the knowledge they have in situations where context is lacking. An alternative way of interpreting these findings would be to take a more positive view and acknowledge the children's number knowledge as reflected in embedded tasks and use this as the basis for teaching them formal code mathematics. This second option seems to me to be of more psychological and educational worth than the first, but it does of course beg the question – 'How?'

In this connection, I followed up an issue raised by Hughes (1983), who found that children who were using written formal code arithmetic in school were still unlikely to recognize a problem as being an arithmetic one in other situations. So, for example, when shown an operation such as two bricks on a table being added to another two and asked to represent the problem on paper, relatively few children produced anything of the form $2+2=4$. None of the children used operator signs and very few even used numerals, leading Hughes to conclude that children do not spontaneously make links between arithmetic in school and arithmetic in everyday life. Davis *et al.* (1985) set out to explore the very basis of what information children need before they will interpret a task as a mathematical one. We began by manipulating the language of the task instructions.

Five-year-old children were shown rows of bricks on separate trials with different numbers of bricks given in each trial. From the child's point of view, we supposed that he or she must make two separate decisions, one concerning what type of information must be represented (about the object, number or both) and a second decision about what strategy might be used to represent that information (to draw, write, use numerals, etc.). We targeted the task instructions to influence these decisions by phrasing the instructions accordingly.

Three groups of children were asked to represent on paper how many were on the table and three groups were asked to represent what was on the table. Within the six groups, the instructions were also balanced as to whether the request to represent was worded as 'write', 'put' or 'show'. We predicted that the instruction 'How many' would lead the children to give information about numerosity, whereas the instruction 'What' would tempt the children towards the objects rather than the numerosity.

The results confirmed our expectations – not only did the 'How many' instruction lead many children to give information about numerosity, the vast majority did so by using numerals alone. In contrast, for those children not given the 'How many' instructions, many simply drew the bricks, often getting the actual number quite correct but not using numerals as a strategy for marking numerosity. Furthermore, we found that in the absence of the 'How many' instruction, children were influenced by whether they were asked to 'write', 'put' or 'show', so that 'write' resulted in attempts at writing, whereas show resulted in a predominance of drawings.

Our study was not designed to test children's mathematical skills but rather to look at how instructions can influence whether or not children interpret a task as being mathematically related or not. The results from our simple study are quite clear: using the mathematical term 'How many' was a powerful force in determining the children's interpretation of what the whole task was about.

Practical implications

I would not wish to argue that task instructions themselves are the only (or even the most important) factor in determining a child's interpretation of a mathematical task. We know from other research that many factors are involved. However, the case for looking more closely at the language used in testing comes primarily from the simple fact that we can control the language we use and be sensitive to the wording of task instructions, whereas other factors such as the child's prior knowledge are less easily controlled. When dealing with mathematically related skills, this argument is particularly apposite because mathematics has its own specific vocabulary which children need to familiarize themselves with if they are to make effective use of the undoubted cognitive skills they already possess on entry into formal education. From this, three key recommendations can be identified:

1. We need to be aware of what we say to children in testing and assessing. As we have seen, minor differences in the wording of tasks can lead to substantial differences in outcomes.
2. We need also to be aware of what we do not say, but which the child might construe. For example, we have seen that the 'mere' fact of repeating a question might signal to the child that we require a new or different answer.
3. We need to embed mathematical experiences in the familiar, but also help children with the 'translation procedures' (Hughes, 1986) from the context-dependent to the formal code.

Conclusion

I hope that the evidence and issues which have been covered in this chapter go some way in suggesting the importance of language in mathematical testing. I have not taken on the issues surrounding whether assessment is a 'good' or 'bad' thing; instead, my concern has been with pulling together the findings from developmental psychologists interested in finding out what and how children understand. The evidence shows that taking language and interaction seriously in assessing knowledge is not just a 'clear up' exercise of improving reliability and validity. Over and above this it leads us to reconsider the overall thinking of young children, which paves the way for theoretical and practical breakthroughs at the research and classroom level.

5 The same number

Richard Cowan

When children answer questions about number in unexpected ways, there are several possibilities. They may differ from adults in any of the following:

1 How they understand the question.
2 What they mean by what they say.
3 The procedure they use to derive their answer.
4 The results they obtain from applying a particular procedure.
5 How they understand number.

The first two possibilities concern the child's understanding and use of language and even such familiar terms as *same* and *more* make demands that are far from trivial. They are examples of the lexical ambiguity abundant in descriptions of number and number relations (see Durkin and Shire, this volume).

Understanding what is meant by *the same*

Same is used to refer to equivalence, but what constitutes equivalence is sometimes left unstated by speakers and so listeners have either to ask for clarification or infer from the context. For example, imagine you are working in a restaurant and in taking the order the customer says 'I want the same wine I had last time.' How do you decide what the customer means?

There are several possibilities: the restaurant may keep unfinished bottles for its customers, and therefore the customer might mean the same bottle. Perhaps the customer means another bottle of the same vintage from the same chateau? Or just another wine made of the same type of grape, e.g. another Riesling? Or just the same colour, such as claret, which used to refer to any light-red wine? Or maybe the customer really wants the same type of bottle and does not really mind what it contains? *Different* is similarly polysemous: imagine what the customer might mean by saying 'I would like to try a different wine to the one I had last time.'

The points I want to make with this example are that difficulties in understanding are not only due to knowledge of possible word meanings and they cannot simply be resolved by looking in a dictionary or paying close attention to utterances. Now you might be tempted to dismiss the problem as one that merely requires a request for clarification, but asking someone what they mean presupposes an awareness of ambiguity and a social context which permits such a request. Where complementary social roles are construed as different in power (e.g. customer/server, teacher/student or adult/child), it may be hard for the less powerful to request clarification without suggesting inadequacy in the utterance of the more powerful. Also such requests may be avoided in conversations between peers because they interrupt the interactional flow.

Developmental psychologists have studied children's understandings of *same* and *different* by asking them to select something from an array of objects that is the same as or different from a target object indicated by an adult. Using such tasks with children between 2 and 6 years of age has yielded the following results:

1 Three-year-olds are just as likely to select a maximally similar object when asked for one which is different as when asked for one which is the same in some way (Donaldson and Wales, 1970; Webb *et al.*, 1974; Glucksberg *et al.*, 1976).
2 Some very young children pick the target object itself as a different one (Webb *et al.*, 1974).
3 Whereas younger children justify picking a blue ring or a white comb as different to a blue comb in terms of similarity, older children mention the difference between their choice and the target (Webb *et al.*, 1974).
4 When asked to pick an object from an array including toy cars and horses that was different from a metal bell, some younger children refuse (Webb *et al.*, 1974).

From such findings, Webb *et al.* (1974) proposed four stages in the acquisition of meanings of same/different:

1 *Different* is synonymous with *same*.
2 *Different* can mean another member of a similar class.
3 *Different* can mean different but there must be some basis of similarity.
4 *Different* can mean completely different.

Glucksberg *et al.* (1976) challenged conclusions about children's knowledge of word meanings drawn from what they do in such tasks. They showed that young children who treat *same* and *different* as though they were synonymous in the selection task context, respond differently when asked for beads of a different colour from when a bead with the same colour is requested. They also created a context in which adults choose a maximally similar object when asked for a different one.

In essence, this controversy touches on a major problem in the study of child language: in theory, it is very easy to distinguish knowledge of word meanings from interpretation of language in particular contexts, but in practice it is very difficult to disentangle them.

Understanding what is meant by *more*

In some situations, *more* refers to the recurrence of an event, in others it means greater in amount. How amount is construed may vary. Consider the problems children confront in learning that one small high denomination coin is considered to be *more money* than a set of several bigger coins.

Across the contexts she studied, Gathercole (1985) found *more* in the greater number sense was generally used in an adult fashion by 5-year-olds. In contrast, *more* meaning greater mass showed marked variation with context: when the amounts differ in mass, children have little difficulty identifying which has more; however, when the amounts are the same mass, children still say one has more.

Furthermore, when the set with greater mass has fewer items, children mostly pick the more numerous set as having more, e.g. when one set is two large pieces of chalk and the other is five much smaller pieces children pick the latter as having more chalk. How to interpret this finding raises the problem of deciding when children's variation from adult language use reflects linguistic or conceptual differences: it could be that the children understand the context as one implying a greater number sense for *more* (linguistic difference) or, perhaps, they genuinely but wrongly think there is more chalk in the greater mass sense in the set of smaller pieces (conceptual difference). The next section discusses some aspects of number development.

Number development

There are many approaches to understanding how children's number concepts develop (e.g. Brainerd, 1979; Bryant, 1974; Fuson, this volume; Gelman, 1982; Gelman and Gallistel, 1978; Klahr, 1984; Klahr and Wallace, 1976; Piaget, 1952).

At present, no approach to understanding children's number development is clearly superior to others in scope, explanatory power or empirical support. Therefore, I shall select points from different approaches which I think a more comprehensive account will have to incorporate.

I shall take the development of understanding of one-to-one correspondence as the focus of number development for children aged 3–7 years. One-to-one correspondence can be defined as 'the process of pairing elements in two sets so that each element in one set is paired with one and only one element in the other set'. Understanding one-to-one correspondence is 'knowing that all those sets and only those sets which can be placed in one-to-one correspondence are cardinally equivalent'. Cardinal equivalence is the most exact meaning of *same number*.

Piaget (1952) identified limitations in children's understandings of one-to-one correspondence from how they performed several tasks including number conservation. In one version of the number conservation task, the adult constructs two sets of items in rows that are equal in number and length. The rows are also made so that each item in one row has a visible counterpart in the other row (perceptual correspondence). Once the child agrees that there are the same number of items in both sets, the adult then transforms the array by either spreading or contracting one row so that the two rows now differ considerably in length. The child is then asked whether the two sets still have the same number. Piaget (1952) classified children's responses to this task in three stages of understanding of one-to-one correspondence. In the first stage, children are unable to construct or even recognize the equivalence of the sets in perceptual correspondence. In the second stage, children change their relative number judgement after the transformation, typically picking the longer row as having more. In the third stage, the initial judgement is maintained.

As Davis explains in Chapter 4, several researchers have challenged the idea that what children do in number conservation tasks is simply determined by their understanding of number. In the present context, it suffices to say that Piaget (1952) highlighted the difficulty young children have ignoring length when it suggests a different relative number judgement to that based on one-to-one correspondence, and he recognized this as a limitation in their understanding.

Brainerd (1973a, b, 1979) assessed children's understanding of one-to-one correspondence from their judgement of static displays in which relative length suggests a wrong answer (see Fig. 5.1). To understand one-to-one correspondence, a child must judge all the displays right without counting. Brainerd (1979) found most children below the age of 11 did not succeed.

So Piaget (1952) and Brainerd (1973a, b, 1979) suggest that learning to trust one-to-one correspondence over relative length takes a long time. This is surprising because one-to-one correspondence is implicit in three common procedures for determining relative number and producing cardinally equivalent sets: counting, sharing and matching. In the following sections, I shall review what is known about how children use these procedures to judge relative number.

Counting

Gelman and Gallistel (1978; see also Fuson, this volume) analyse counting in terms of three how to count principles;

1 *One-to-one*: every item to be counted should be tagged with one and only one tag.
2 *Stable order*: the order of tags must be the same in every count.
3 *Cardinal*: the last tag used is the symbol for the number of items counted.

The first principle is a form of one-to-one correspondence and several studies suggest children understand this and the other principles of counting from an early age (Briars and Siegler, 1984; Gelman and Gallistel, 1978; Gelman and

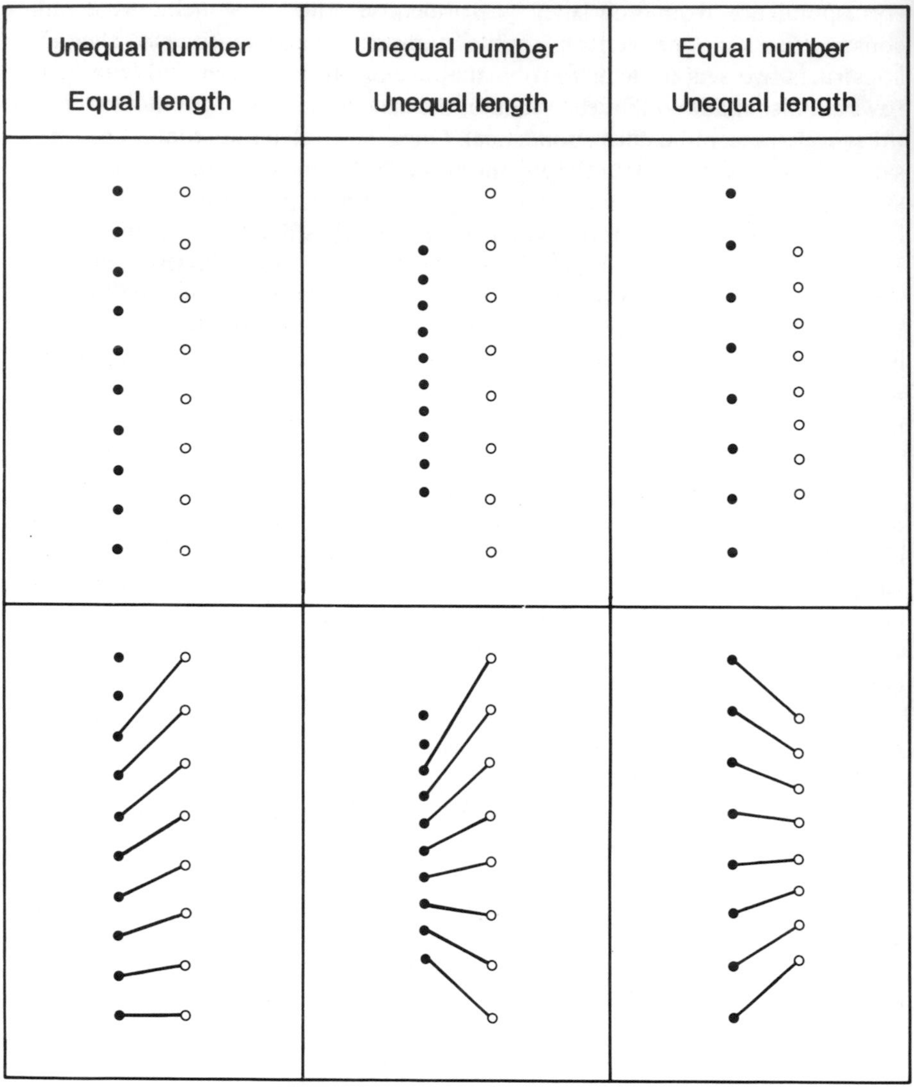

Unequal number Equal length	Unequal number Unequal length	Equal number Unequal length

Fig. 5.1. Display types used to assess children's understanding of one-to-one correspondence by Brainerd (1979) and guidelines versions used by Cowan (1987b) and Cowan and Daniels (1989).

Meck, 1983). On the other hand, we also know that children do not use counting to solve number problems (Michie, 1984; Sophian, 1987). For example, few 3- and 4-year-olds used counting to make a set of objects that was the same number as another set (Sophian, 1987).

I wanted to find out when children would use count information to judge

ative number (Cowan, 1987a). I asked them to judge rows of blue and yellow ɔts arranged in displays. The types of displays comprised Brainerd's (see Fig. .1) and length-consistent ones with small and large number versions. The small number versions had rows of 3 or 4 dots. The large number versions were within the children's counting range and had rows of 8 and 9 dots or rows of 15 and 16 dots. Some children had to count the dots themselves, whereas for others I did the counting.

Nursery school children, aged 3–5, typically judged the length-consistent displays to be correct, and overall it did not matter who counted. Brainerd's displays were frequently judged wrong, especially the large number versions. The 5-year-olds differed only in that they rarely misjudged the small number versions of Brainerd's displays. Only with 6-year-olds were errors in judging the larger number versions of these displays rare.

I take these results as evidence that the understanding of one-to-one correspondence is slow to develop, but there are four other explanations of why children in general might not use count information to judge relative number. I shall briefly describe these and discuss whether they could account for my results. The first is that children might simply not be able to count properly; a child who does not know the cardinal principle, as described by Gelman and Gallistel (1978), can hardly be expected to use counting to compare. But in my study, only children who passed a counting test which involved the cardinal principle took part.

The second explanation is that children do not know how to adapt counting to the task. Saxe (1977) found most 3-year-olds simply counted on when asked to compare the numbers of pigs and horses. However, I told children who counted for themselves to first count the blue dots and then the yellow dots. They all did so.

The third explanation is that children might lack confidence in their counting. When the counting is done by an adult, it might be regarded as more reliable. But in my study, children judged no better when I did the counting for them. Finally, memory failure might explain why children do not base their judgements on counting; children may forget how many there are in the first set by the time they finish counting the second. I do not think this explains my results, as I reminded children how many there were in each set just before they were asked to judge.

Therefore, it does appear that trusting counting over length appears several years after learning to count. Gelman (1982) sees this as a matter of the child learning to access the one-to-one correspondence implicit in counting (the one–one principle), and she suggests that such access can be obtained by counting small sets of items before and after transformations. The size of set is important because it is in counting small numbers that children first become reliable and confident (Gelman, 1972). Gelman (1982) found a brief training session, in which 3- and 4-year-olds counted rows of items before and after transformations, was enough to produce number conservation on both small and large number versions.

Experience with small number sets is considered to play an important part in children's number development by Klahr (1984; Klahr and Wallace, 1976). However, unlike Gelman (1982), Klahr suggests subitizing rather than counting is the crucial procedure. Subitizing is a process for determining number, which in adults is very rapid and limited to numbers less than eight. They propose that subitizing develops before counting, because, among other reasons, it does not depend on verbal instruction. Because children's subitizing range is even more limited than adults, then number sets have to be very small, (i.e. 2 or 3) for children to subitize them.

Sharing

Sharing a set of objects by allocating them one at a time to each individual, which Miller (1984) calls 'distributive counting', embodies the principle of one-to-one correspondence. Two studies have shown 3-year-olds to be remarkably competent in sharing objects between two or more dolls or turtles (Desforges and Desforges, 1980; Miller, 1984). But Frydman and Bryant (1988) pointed out that the children could have simply been using a procedure which they had learned without understanding. They assessed children's understanding of sharing in two ways; first, they told children how many items were in one of two equal shares and asked them how many were in the other and, secondly, they asked them to share objects that were either singles or stuck together in pairs or triples to see if the children would adapt their sharing. The first task proved to be difficult for the 4-year-olds who took part; none saw there was no need to count and all of them tried to. When this was precluded by covering the objects only 10 of the 24 children answered correctly. Very few 4-year-olds adapted their sharing, but most 5-year-olds did. However, Frydman and Bryant (1988) report a further study which showed it was simple to train 4-year-olds to adapt their sharing.

In another study (Cowan and Biddle, 1989), children watched objects being shared and then judged whether the shares were fair. Whether the shares were equal was varied, as well as whether the shares could be seen when they were judged and the number and nature of objects shared. Sometimes, an object was left unallotted. Overall, the results suggested the children, between $3\frac{1}{2}$ and $4\frac{1}{2}$ years old, understood the principle of one-to-one correspondence but were limited to applying it to small numbers.

What is unknown is when children will trust sharing over relative length. Piaget (1952) described children who abandoned their belief in the equality of shares when the elements of one share were spread out or made to form a more compact heap. This has not been followed up.

Matching

Bryant (1974) discussed matching as a fundamental procedure embodying one-to-one correspondence. By matching, he means pairing off items in two sets so

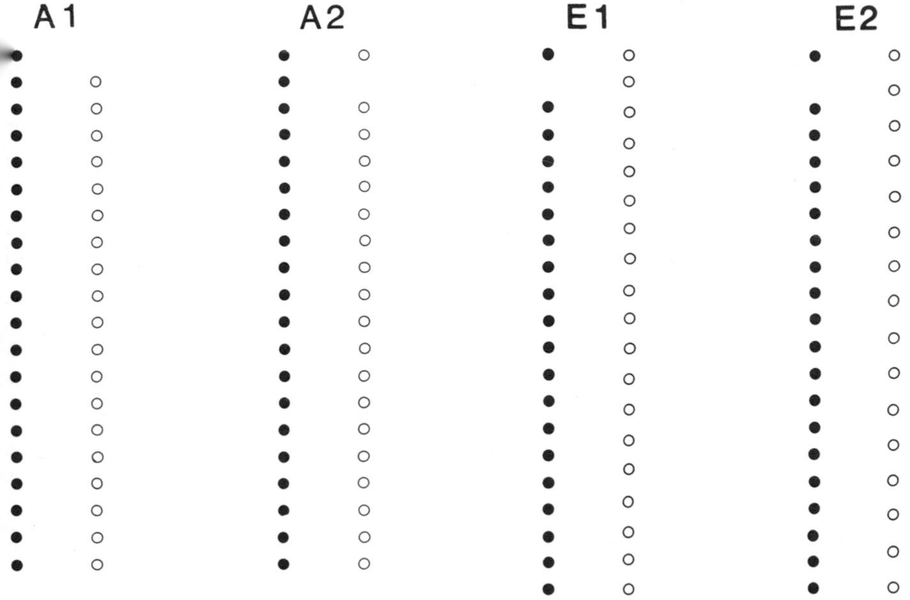

Fig. 5.2. Displays used in studies of children's relative number judgements. Bryant (1974) suggested matching was the basis of children's success in judging A-type displays. Cowan (1984) suggested children simply judge A2 displays on the basis of the gap which is why they misjudge E-type displays.

that each item in one set has one and only one partner in the other. He proposed that matching was the basis on which children as young as 3 successfully judged A-type displays (see Fig. 5.2). By pairing off the items in the two rows, they discover that one row has an item which cannot be paired off and therefore conclude that that row has more.

There are some problems with using A-type displays as tests of children's understanding of one-to-one correspondence (Cowan, 1984). One problem is that in A1 displays, the column with more dots is also longer and so correct judgements may be based on relative length. Also, whether 5-year-olds do necessarily judge A2 displays by matching is doubtful: seeing only one column at a time prevents matching but does not cause them to judge any worse, and they are just as likely to judge A2-type displays right as they are to judge E-type displays (see Fig. 5.2) wrong (Cowan, 1984). The latter suggests they are simply judging on the basis of the gap.

Three methods have been used to encourage children to use matching: using complementary sets of objects, e.g. eggs and eggcups (provoked correspondence); using heterogeneous sets of objects with corresponding partners in each set (unique pairs); and drawing guidelines between the members of each set.

When number conservation is the measure of understanding of one-to-one correspondence, the results are mixed. Dodwell (1960) found provoked correspondence helped children conserve, but Miller and West (1976) identified methodological problems in this and other early studies. Miller *et al.* (1975) found unique pairs helped; however, Whiteman and Piesach (1970) found children only benefited when the unique pairs were joined by guidelines. In both studies, there were only four or five objects in a set. Miller and West (1976) found no method helped with sets of seven objects, but Fuson *et al.* (1983) found children between $4\frac{1}{2}$ and $5\frac{1}{2}$ were much more likely to conserve when the seven toy animals and peanuts were tied together and the connection between animal and nut pointed out.

Other studies have been carried out using judgement of static displays. As mentioned earlier, Brainerd (1973a, b, 1979) found most children below the age of 11 failed to judge all his displays (see Fig. 5.1) right when they were not allowed to count. But perhaps it was not their lack of understanding of one-to-one correspondence that was responsible. Instead, children may decide that matching is too difficult to apply to these displays (Avesar and Dickerson, 1987), or they may try to match but fail to pair off items in the two rows correctly. When guidelines are drawn between the dots in the two rows (as in Fig. 5.1), children must still judge contrary to relative length to succeed, and thereby demonstrate their understanding of one-to-one correspondence, but the practical difficulty in pairing off items is much reduced. Adding guidelines helped both 5- and 7-year-olds, but whereas 5-year-olds still typically judged wrong, 7-year-olds rarely erred (Cowan, 1987b).

Similarly, guidelines helped children judge different number versions of Brainerd's displays (Cowan and Daniels, 1989). In both these studies, judgement accuracy varied with display type. Guidelines have also been found to help 4- and 5-year-olds judge a variety of other display types (Avesar and Dickerson, 1987; Cowan, 1984). When small and large number versions of displays with guidelines are compared, there are always some children who judge the smaller number versions better.

Overall, providing guidelines is the most effective way of enabling children to use matching, but even with guidelines children less than 7 years of age are likely to misjudge relative number when relative length suggests a wrong judgement and the numbers are large.

Practical implications

1 As in many other aspects of language use in mathematical work, care must be taken when using terms such as *same*, *different* and *more* and interpreting children's use of them. Their meaning varies with context and can pose problems, even in interactions between adults. In interacting with children we cannot assume they will understand what we mean or ask us to explain.

2 Knowing how to count, share and match is an important part of number development. Children need practice with these procedures so that they can use them reliably and confidently. They also need to be taught how to use them to judge relative number.
3 Knowing the procedures is not the same as understanding them. That one-to-one correspondence is the definitive criterion for cardinal equivalence is not self-evident. Children need to learn that it is superior to relative length and why. The way children disregard relative length with some displays before others and with small number versions before larger ones suggests that, currently, such learning is gradual rather than sudden.
4 Children do not automatically generalize their understandings from small to large numbers. They need to be helped to do so.

Conclusions

In most of the studies I have described, the goal has been to improve our understanding of how children's performance of number tasks varies under different conditions. In contrast, studies which have deliberately tried to change children's performance are rare. However, these may be more useful for teachers, who are often more concerned with improving a child's understanding than with describing it. The exceptions are the studies by Gelman (1982) and Frydman and Bryant (1988). In both of these studies, brief sessions sufficed to produce major changes. Without evidence it would be foolish to conclude that all defects in children's understandings of number can be so swiftly remedied, but equally without conducting training studies it would be wrong to infer from studies showing a protracted development that such gradual change is either inevitable or desirable.

Finally, there is still much we do not know about children's number development even in the limited aspects I have discussed. For example, we do not know when and how children see the relation between counting, sharing and matching, or whether children transfer their knowledge of the unreliability of relative length from one procedure to another. Also, it is always possible that what we believe we know can be better understood in a fundamentally different way: perhaps the use of relative length reflects a different understanding of how accurately we want relative number to be judged. After all, comparing lengths is a quick and easy way of estimating relative number and such a rough guide can be adequate for various purposes. Perhaps the significant changes in the early school years are in understanding and identifying different purposes underlying school tasks, but children must at some point acquire the numerical skills and understanding necessary to perform them. I believe these are acquired by many only when they have started school, for I have often observed children who are amazed and excited to discover from counting and recounting Brainerd's *Equal number/Unequal length* display that they 'really are the same number'.

Further reading

Bryant, P.E. (1974). *Perception and understanding in young children: An experimental approach*. London: Methuen.

Gelman, R. and Gallistel, C.R. (1978). *The child's understanding of number*. Cambridge, Mass.: Harvard University Press.

Ginsburg, H.P. (1977). *Children's arithmetic: The learning process*. New York: Van Nostrand.

Hughes, M. (1986). *Children and number: Difficulties in learning mathematics*. Oxford: Blackwell.

Piaget, J. (1952). *The child's conception of number*. London: Routledge and Kegan Paul.

6 Children's production and comprehension of written numerical representations

Anne Sinclair

Our written numerical system is simple, precise and the best tool we possess both for computation and the construction of mathematics. However, it took humanity several thousands of years to construct it, i.e. to systematize and make full use of both zero and place value (Ifrah, 1987; Kearins, this volume). As both common sense and previous research shows, these two indissociable characteristics (zero and place value) are difficult for children to grasp; before fully understanding them (and thus having some grasp of the entire system), they understand other, simpler elements of the system. For example, typical 5-year-olds will readily and correctly write 3 when asked to 'write a three'; they may correctly write 5 when asked, for example, to write down how many chips there are on the table and when the number exceeds ten, they 'give up'. Similarly, when asked to 'read' (explain, interpret) a notation like 45, they may say 'a four and five, I don't know how much that makes!', seeing the system as an additive one, which is correct, but ignoring place value, which is multiplicative. Older children (Kamii, 1980) may correctly write 16 of the sixteen chips they have just counted, but they will not be able to establish correspondences between the 1 and the 6 in the notation and the collection of chips in front of them. One may give similar examples for the difficulty of comprehending and producing the various uses of zero in our system: in a taking-away task, many 4- and 5-year-olds will say 'then there's nothing, I can't write that down' for empty sets, or, when asked to 'write a ten', 'Ten, I know that one, that's one and a letter O', etc. Undoubtedly, grasping our numerical written system (whole numbers only) is a development that takes several years.

In this chapter I shall discuss the very beginnings of learning to write (and interpret) numerals, i.e. the gradual understanding of some of the simpler aspects of the system. I shall thus concentrate on the written representation of cardinalities less than ten as carried out by young children, and shall briefly describe some data on how young children interpret environmental written numerals. The data discussed come from studies carried out with children aged

3–6 (preschoolers in kindegarten or day care in Geneva) – an age range where most children are not yet tackling zero and place value as problems to be solved (although they may already have intuitions and ideas about the observable results of these characteristics, such as for example, the idea that 'more numbers is more').

Early attempts at written numbers

What do children do when asked to represent ('note', 'mark' or 'write down', following the child's own vocabulary) sets – cardinalities one to eight – of identical objects laid out before them? Most children aged 3–5, as well as some 6-year-olds, use one-to-one correspondence procedures. (A few children aged 3–5 will exhibit more sophisticated procedures, and a few others will give responses of a different type, such as for example, attempting to write 'stars' with letters: see Sinclair *et al.*, 1983; Sinclair and Sinclair, 1986; Sinclair 1988.) Similar behaviours have been reported by Allardice (1977), Hughes (1982) and Sastre and Moreno (1976) with, respectively, American, Italian and Spanish children. These one-to-one notations may be of various types, but they are almost always correct: only a few children aged 3 make occasional errors of one element.

Some children use the same graphic mark throughout, e.g. a short bar (tally mark), a cross, or some other simple shape, using the same grapheme to represent sets of pencils, houses, chips, small balls, etc. Others will use simple shapes – perhaps iconic in nature, but nevertheless very different from drawings – of different types for different kinds of objects. For example, one child I interviewed used open round marks to represent poker chips, strokes to represent pencils, and potato-like shapes to represent small houses.

Many children use letters or mock letters (graphemes that resemble letters in their arrangements of curved and straight lines); those who use letters use them throughout. Various patterns appear: some use their (often small) stock of shapes to make varied sequences (e.g. IOUP for four houses, PIOU for four balls, IPI for three houses, etc.); others prefer to align identical shapes for one display (e.g. AAA for three houses, ppppp for five chips, etc.).

Some children use numbers; a few write down the sequence 1234 . . . etc., corresponding to the number of objects (e.g. 12345 for five chips), and others write the cardinal number down the number of times that is correct according to one-to-one correspondence (e.g. 4444 for four houses). Figure 6.1 gives some examples of one-to-one correspondence notations, for a display of five small rubber balls, not linearly arranged.

The children consider that their notations represent the displays in their entirety. This is obvious from the fact that they do not try (or take up the suggestion) to write down the name of the objects represented and from the way they re-read their notations. When asked 'what did you write down?', 'explain to me what is written here?' or 'tell me again what that says?', they respond with 'those' (or equivalent expressions, pointing to the display), 'chips', or they

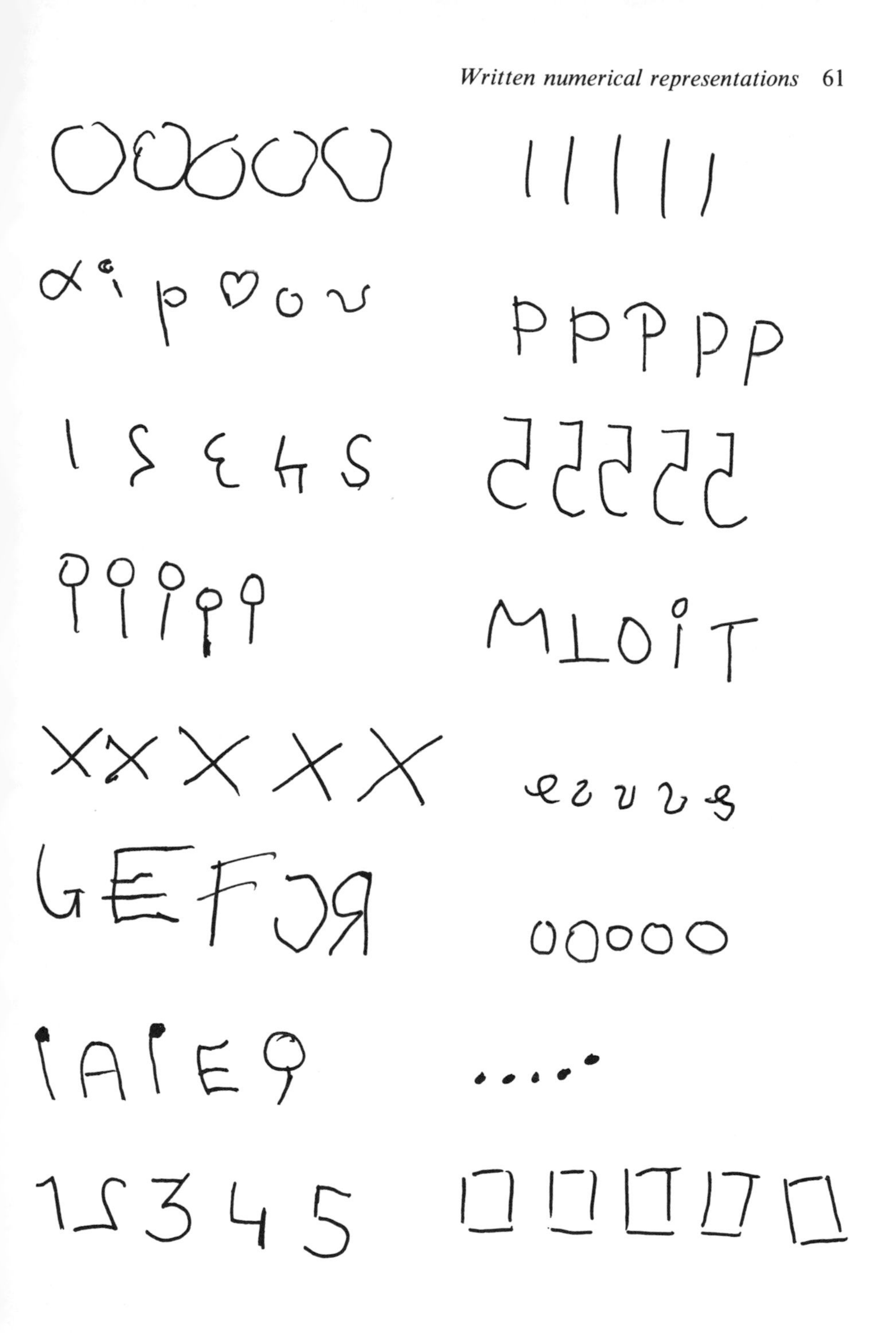

Fig. 6.1. Some examples of one-to-one correspondence notations for a display of five small balls.

respond with a spoken numeral and the name of the object, as in 'it says three houses'.

With one exception (see below), it is difficult to relate these behaviours to age or to performance on other tasks dealing with number. We found that, overall, the type of graphic forms used in one-to-one notations is not age-related: some older children use tally marks, some younger children use mock letters, etc. The forms used appear also to be unrelated to counting capacity or knowledge of number shapes. Naturally, 3- and 4-year-olds exhibit primitive counting behaviours; at age 5–6, counting strategies improve, many of our subjects being capable of correctly and rapidly counting the arrays we present. Yet all these children produce essentially the same type of notations. The exception is that the production of written numerals is age-related; it increases with age.

Continuities and changes in written numbers in the kindergarten years

Development continues as follows. A few 5-year-olds and many 6-year-olds write down one numeral only, and some add letters or marks to signify the name of the object, thus behaving in the same way as adults do (e.g. 4 mézone for four houses, 5 K for five crayons).

These data deserve some comment. First, it is clear that representing quantities with paper and pencil is easy even for very young children; secondly, the use of one-to-one correspondence procedures develops early and this behaviour persists over a large developmental span (the difference in conceptual development between 3½-year-olds and 6-year-olds is considerable).

That these notations are developed early (and I have seen 2-year-olds use it) is not surprising: babies carry out one-to-one correspondence procedures in action from the age of 22 months (Sinclair *et al.*, in press), e.g. putting a pencil in each cup (with six pencils and six cups), a block on each car, etc. Making a mark for each object is not so very different; one must be able to hold a pencil, and there is a small measure of spatial dissociation between the two terms (object–mark).

That one-to-one notations persist for so long, and that they are used by children who are capable of writing the numerals 1–9 (when simply asked 'Could you write a five for me please?') is perhaps more surprising. Yet, one-to-one correspondence is a very powerful scheme that will be used throughout life to handle tasks and solve problems; it is the cornerstone of any kind of numerical representation (mental, verbal or notational); and in any system, one entity must at the same time be made to *correspond to* and *stand for* a particular (or indefinite) object or set of objects. Furthermore, the oldest numerical notational systems uncovered (Upper Paleolithic and Neolithic) are of this type – tally marks without special signs for powers of a base (see Struik, 1967; Guitel, 1975; Marshack, 1972; Ifrah, 1987). Viewed in this light, it is perhaps not so surprising that one-to-one correspondence procedures are so pervasive in children's numerical notations.

Cardinality in written and spoken numerical representation

These results also suggest, however, that despite good counting capacity and a good knowledge of number shapes and their names (many subjects using one-to-one notational procedures correctly produce number and letter shapes on demand), the exact meaning of a written numeral is not understood until later (about age 6 in our study). Specifically, their cardinal value (see Fuson, this volume) is not understood. They are either used as tags for particular individual objects, or they are intimately associated with the oral counting sequence, i.e. their ordinal aspect is at least partially understood, but their cardinal value is not. We thus see that the development of numerical notation is highly dependent on the development of number concepts; 'knowing how to write a five' (and all other numerals) has but little to do with the appropriate use of written numerals and their integration in a system.

Research carried out by Piaget's collaborators clearly shows that understanding the cardinal value of *spoken* numerals is difficult for children. Gréco (1962), for example, working on the links between counting and number conservation (in the classical task of spatial displacement of a row of chips), showed that many children aged 4½–7 fail number conservation (saying there are more chips in one or the other line, with the traditional unshakeable arguments) but correctly predict the number of chips in the line without counting, thus using a correct reasoning based on a previous count. For example, Gréco (1962: 7, my translation) quotes the following child who responded to the conservation question by stating that there were more chips in the blue line of chips which was spread out:

Experimenter: How many reds?
Child: Six.
Experimenter: And how many blues?
Child: [immediately] . . . Six.
Experimenter: And you said there were more where?
Child: If you count, each of them is six, but if you don't count, there's more blue ones, it's long, long . . .

The following is an example where the conservation question was on the relation between two unequal collections (the child noted the lack of equality in the one-to-one correspondence configuration), i.e. seven blue chips spread out *vs* eight red chips close together (Gréco, 1962: 7, my translation):

Child: Here [blue] there's more, it's longer.
Experimenter: How many reds?
Child: Eight [correct].
Experimenter: And how many blues, without counting?
Child: Seven [correct], I counted before.
Experimenter: So, are there more red chips or more blue chips?
Child: Before, more red chips, and now more blue chips.
Experimenter: How many blue chips now?

Child:	Seven.
Experimenter:	And how many red chips?
Child:	Eight.
Experimenter:	More blue chips or red chips?
Child:	More blue chips, *voyons!*
Experimenter:	What do you think? What is more, seven or eight?
Child:	Eight of course!
Experimenter:	So . . . but here, seven blue chips is more than eight red chips?
Child:	No – eight is more than seven, but there, there are more red chips!

No-one who has used the number conservation task will be surprised by these examples. Gréco's figures for conservation of number in the various (rather difficult) tasks he presented range from 3% at age 4½ to 40% at age 7, with the following proportions of children making correct *numerical* predictions of equivalence or difference: age 4½, 10%; age 5, 30%; age 5½, 43%; age 6, 69%; age 6½, 73%; age 7, 63%. The cardinal value of the spoken numeral (as the result of a count mentally kept track of) is not robust enough for many children to resist a non-conservation judgement.

Thus, on a conceptual level, cardinal value is grasped at a relatively late age, despite 'correct' counting. We should not be surprised, therefore, that in our much simpler task (no comparisons are involved and arrays are presented as static, and the child can manipulate them freely), written numerals used for their cardinal value do not appear before the age of 6.

Written numbers in children's everyday environments

If we turn our minds to the young child's everyday experiences with written numerals, other considerations support the lack of salience of cardinality. Before entering school, where the child will meet written numerals representing cardinal numbers in simple counts or equations (e.g. $2+2=4$), the child will not see or hear discussed many (if any) written numerals of this type (unless he or she is exposed to someone who behaves like a schoolteacher, getting the child 'ready' for school). The written numerals present in the environment that may solicit ideas or hypotheses, and that may be discussed and commented on by other people, include speed limits, bus numbers, page numbers, house numbers, telephone numbers, dates, etc. Many of these numerals are ordinals, e.g. page numbers, house numbers, winners of races and, very important for young children in my own community at least, the numbers of lift buttons. Many others are merely tags, e.g. bus numbers, telephone numbers and TV channel numbers. Yet others are measures of some kind, e.g. time, speed, weight and distance. These quantitative measurements, though they include the cardinal aspect of our numeration system, are notoriously difficult for children to grasp (often also for adults), as they are based on arbitrary points on a continuum, or even relations between continua. Prices are somewhat similar and yet different.

Other numerals are 'measures' or 'classifications' which are hard to define (it is often difficult even for adults to explain their exact nature), but they certainly do not deal with cardinality, e.g 18k on gold ring, 60 on a light bulb, 1 on the carriage of a train, 25 on the sole of a shoe, etc. This, then, is the material children are presented with: only a few exceptions come to mind. Though they see cardinals on the packages of a few manufactured goods (e.g. 20 super CHOCOS), such numerals are rare. The only equation they will ever see before going to school is a receipt for groceries. The fact is that outside the school setting, and the domains of arithmetic and mathematics, written numerals in our society only rarely represent cardinalities (spoken numbers are another matter).

When one asks children to interpret ('tell me what you think it says, what it means, what it tells the people who look at it, etc.') environmental numbers, such as those mentioned above, children unsurprisingly enough do not give interpretations based on cardinality, but interpretations based on other aspects, irrespective of the fact whether they can 'read' (identify, name) the numeral or not. For example, 'classification' or 'grouping' responses for the number 22 on a bus include:

It tells which kind of bus it is.

It tells which number it is, where it goes.

That it is a bus and not a truck.

Others are what one might call 'tagging' or 'correspondence' responses, e.g. for a car number plate:

So that the police knows it is your car, not somebody else.

So you know where to park.

Order is also salient:

If you have to go forwards or backwards [for a numeral on a house].

Who goes first and who goes second [for numerals on running vests].

Responses of a cardinal nature ('how many . . .') or of a clearly quantitative type ('how much . . .') are rare, but this naturally depends in part on the items one presents (see Sinclair and Sinclair, 1984).

In our notational tasks, the use of counting principles, or what one might call pre-counting principles (one-to-one correspondence, iteration, listing, enumeration, partial orderings) and verbal counting, whether adequate or not, is very evident, both in the procedures used and in their results, (e.g. touching each chip before marking, writing down the counting number, etc.). Indeed, while data from the area of conceptual development may help to explain why young children did not use cardinal numbers in our task until the age of 6, it is via counting that they find the task easy and perform it well, consistent with Fuson's (this volume) results on the crucial contributions of counting to children's discoveries about number.

Further development

Once children have understood that our written numeral system, as used in primary arithmetic, represents cardinality (i.e. the notation 9 generally stands for a set of nine, abstract or concrete), they must still puzzle out place value (and the use of zero). Several different strands of reflections seem to play a role (e.g. children talking about one or two 2-digit numerals, and corresponding collections of objects, in ongoing research). Ideas about addition ('the two numbers have to be taken together', 'you add the two', 'more bigger numbers is more', etc.), partition ('one is for a part, the other for another part'), order ('not the same, here the five is in front, there it's in back'), position ('the first number counts for more'), multiplication (for 64: 'this one [6] will be for six times four I suppose') and about the zero ('zero is for big numbers') will all be important. As several previous studies have shown (e.g. Kamii, 1980), place value is not really grasped before the age of 8–9, or even later.

Practical implications

What are the implications for education? In areas like the reading and writing of numerals, children's knowledge develops thanks to information provided by the environment and according to how exactly children 'read in' and integrate that information. Probably because of this fact, studies such as those discussed here do not uncover general, profound, cognitive developmental principles, nor anything resembling 'stages'. If one believes that culture, education and cognition interact, there seems little point in discussing the role such results should play in the setting up of education guidelines or programmes.

Results such as those presented here are nevertheless of potential use to the teaching community: they may serve to increase teachers' knowledge and understanding of the children they teach. In most Western school systems, teachers are well aware of the fact that beginning first-graders already know 'something' – or 'quite a lot' – about written numerals. Often, too, they will see children's knowledge as fragmented, sketchy or even 'disorganized' rather than merely partial. What they often do not realize, however, is that much of the children's seemingly fragmented knowledge may be based on or linked to profound ideas that are coherent, if erroneous from an adult point of view. A child who 'spontaneously' does additions by writing down 333 and 22 and counting all the numerals and then writing 55555 in one-to-one correspondence we met several such children in our studies), will have particular difficulties in understanding equations like 2+2=4.

Many children beginning school do not yet fully grasp the cardinal value of figures like 5 in isolation. Their first ideas when faced with the shape 5 in some written context may be the association with the tag 'a five' (which is different from calling it 'five' and thinking of a cardinality) and possibly a search for contextual meaning that does not have anything to do with cardinality and computation. They may thus learn simple additions and subtractions by rote; i.e. 'knowing' some $n+2$ equations and not others, not yet having developed a 'counting-on' strategy to solve such problems. For example, when a 5-year-old was asked to complete the equation $2+2=$, she wrote a clumsy four, and (under pressure) explained: 'Two and two makes four . . .? Look, [points to four virtual ordered imaginary spots on the table] two, two, two, two [returns to the start and counts the spots]. One, two, three, four'. Is rote learning (and its attendant sometimes peculiar integration) necessarily negative? One cannot possibly condemn 'rote learning' out of hand (and I cannot discuss such large questions here), but such children's judgement of what is *arbitrary* (number names, number sequence, number shapes, direction of writing, etc.) and what is *not* (e.g. link between sum and result, commutativity of addition, etc.) will be completely misguided. This is surely not a good start, even if such thinking is rapidly revised.

One point of a general nature arises. It seems that some children's difficulties are in part due to the fact that we try to teach children to *use* written symbolic systems appropriately (this *is* an all-important outcome of education) without ever specifically *explaining* these systems to them. For example, we (parents, teachers, etc.) do not tell children that written numerals do not represent sound or spoken numbers, and that one cannot expect (in English) to find any reasonable, consistent link(s) between spoken numbers and written numerals. But we do tell them, for example, that 'one and one [11] is eleven', thereby implying the opposite (new symbol = different word). However, for most young children (even those who can count very well), the numerals 11 can only mean 'one' (that's what's written there, after all, even if it is written twice) 'two' (there are two figures, or one and one is two), or possibly 'more than the numbers written singly', i.e. 'ten', 'twenty', 'a hundred', 'a lot', etc.

Conclusions

In this chapter, I have described some results concerning young children's production of numerical notations, both conventional and unconventional, and I have briefly touched on their understanding of some environmental written numerals. I have linked these results to, or interpreted them in terms of, the construction of the cognitive concepts (principally cardinality) that underpin our notational system. Such attempts are often misunderstood, and are seen as propounding the view that the mysterious, internal process that psychologists call 'cognitive development' is ultimately responsible for educational achievement (particularly in the primary years and in areas such as symbol systems, arithmetic, problem solving, etc.). Cognitive development takes place in a cultural, educational setting (cf. the chapters by Clarkson and Kearins, this volume), and even a content-free, most likely universal achievement such as number conservation admits to *very* considerable environmental differences (e.g. Lancy, 1983).

It is obvious to me that children construct an understanding of cardinality (which will permit them to grasp certain aspects of our written numerical system) not only through abstractions from their own actions (*à la* Piaget), but also through contact with counting, quantification, written numerals, etc., in interaction with other people. Carrying out activities where these play a role (playing games, solving problems, doing school work, etc.) will be central. However, the internal interplay of all these elements remains largely mysterious.

Acknowledgement

The author gratefully acknowledges the support of the Fonds National Suisse de la Recherche Scientifique, grant no. 11–25427.88.

Further reading

Hiebert, J. (1988). A theory of developing competence with mathematical symbols. *Educational Studies in Mathematics*, **19**, 33–355.

Hughes, M. (1986). *Children and number: Difficulties in learning mathematics*. Oxford: Blackwell.

Section 3

Language and meanings in mathematical education

7 Lexical ambiguity in mathematical contexts

Kevin Durkin and Beatrice Shire

Teacher: Let *n* be a number.
Pupil: But Miss, *n* is a letter, not a number.
(after Adda, 1982: 211)

The Mad Hatter insisted that words were to mean what he wished them to mean. In mathematics education, people appear to do this kind of thing frequently, and guests at this party may well find that many of the assumptions that they hold about meanings of words do not withstand the unpredictable verbal transformations wrought magically by their hosts. Like Alice, they may find that nothing is quite what it seems.

This chapter is concerned with some of the lexical (i.e. word) ambiguities that children are likely to encounter in the language of maths. Lexical ambiguities do not exhaust the variations of linguistic ambiguities that pupils face in mathematics, and it should be clear that our focus is on only one aspect of the linguistic challenge of maths education. However, we will attempt to demonstrate that this is a diverse and widespread challenge, involving a wide range of terms. First, we review the kinds of lexical ambiguity that the language of school maths entails; many other authors have noted these ambiguities, and we draw on earlier research and anecdotal evidence which indicates that lexical ambiguity does confuse pupils. We turn then to a particular subset of ambiguous words in maths, namely spatial and size words (such as *high* and *big*) and summarize some of our own research into young children's interpretations of these words when applied to numbers. We consider next some of the ways in which children are likely to experience uses of ambiguous words in maths education, including classroom interaction, maths broadcasts and written materials, and propose that the difficulties that arise may contribute to poor performance and pupil anxiety in mathematics. Finally, we consider the practical implications and the prospects for intervention.

Types of lexical ambiguity in mathematics education

Linguists have identified several different types of lexical ambiguity, including homonymy, polysemy, homophony and shifts of application (cf. Bennett, 1975;

Lyons, 1977; Panman, 1982; Deane, 1988). In this chapter, we will be concerned principally with homonymy and polysemy, but to set these in context it will be useful to note the various types of ambiguity, as each is relevant to language in mathematical education.

Homonymy denotes the property of some words that they share the same form but distinct meanings. A standard example is *bank*, meaning financial institution, and *bank* meaning an area of land beside a river. An example relevant to mathematics is *leaves*, which in everyday use can refer to the outgrowths of a tree but which has a quite different meaning when it refers to the process of subtraction ('3 from 7 leaves 4').

Polysemy refers to the property of some words that they can have two or more different but related meanings. The word *mouth* presents a standard example of polysemy; this word can refer to the principal facial aperture or to the place where a river meets an ocean. These are different meanings, but it is apparent to the native speaker that they contain some shared sense (in this case, 'opening'). A mathematical example would be *product*, which in everyday use refers to a thing which has been made, whereas in maths it refers to a quantity obtained by multiplication. Again, it seems obvious that there is a shared sense of 'that which is produced' in both the everyday and the specialized meanings.

Homophony refers to the phenomenon wherein two distinct words have the same pronunciation. Standard examples include *bare/bear*, *flour/flower*. These are linguistic coincidences which might lead to occasional misunderstandings and are frequently employed in puns. Mathematically relevant examples include *two/too/to, four/for, sum/some, pi/pie.*

Shifts of application refers to occasions where the same sense can be considered from different perspectives. An example from Grober (1976) is *wall*, which has different aspects according to whether it is discussed with reference to its composition (e.g. brick, stone) or its function (e.g. from the perspective of bricklayers, architects, residents). The word *number* is subject to shifts of application, for example when used to describe nominal ('the number 5'), ordinal ('the second number she said'), cardinal ('the number you counted') or visual ('the number 7 is crooked') properties.

We could add a further category of ambiguity, that of imprecision or sloppiness on the part of the language user. This could occur, of course, in any area of vocabulary use, but we should note in passing that it is frequent in mathematical contexts. Some examples may be relatively minor, such as the use of *average* when *mean* is intended (which the average statistical textbook will tell you is undesirable), or the interpretation of *data* to indicate a singular referent. Others may be more misleading, such as the use of *share* to refer to division (which works satisfactorily for some problems but less well for others: see Anghileri, this volume; Hanley, 1978) or *diamond shape* to refer to a square in a particular orientation (see Hanley, 1978: 28). Certain words are prone to be used informally in ways which conflict with their precise meanings in mathematics and logic, such as the quantifiers *some*, *any* (see Pimm, 1987: 79) or *more* (see Cowan, this volume). For example, Hobart (1980) notes that adults might say 'Let's move the tables to make more space', a description which would fail a Piagetian conservation of area assessment!

It is important to be aware that most language use involves shifts of application (otherwise we would never say anything new), and that even maths educators who pride themselves on the precision of their subject are as capable as any other language users of imprecise and misleading descriptions (some examples are discussed in the chapters by Pimm, Kerslake, Nolder and Anghileri in this volume). These issues raise practical and research questions for mathematics education. Our focus here will be mainly on ambiguity which is systematic in that the multiple senses of the words are relatively consistent. That is, we will be concerned with the relations between the basic, everyday meanings of words such as *leaves*, *product*, *tables*, *high*, *big*, and their specialized uses in mathematical contexts. Thus, most of the ambiguous words we will discuss are either homonymous or polysemous. These types of ambiguity are of interest because of the possibility that we can identify the basis for particular misinterpretations by pupils, and hence develop teaching strategies that circumvent or exploit such tendencies.

Examples of ambiguous words commonly used in school mathematics

In addition to the individual common ambiguous words used in mathematics listed in Table 7.1, certain word *combinations* created within mathematics are intended to represent meanings which are different from the sum of their everyday senses, e.g. *simple interest*, *pie chart*, *square root*, *closed figure* (Shuard and Rothery, 1984: 28). Conversely, some colloquial descriptions of mathematical operations have alternate everyday meanings, such as *take away* (*takeaway*).

Some lexical ambiguities are compounded by the fact that they relate to mathematical symbols which themselves are described by different words in different contexts. For example, '=' can mean *equals*, *means*, *makes*, *leaves*, *the same as*, *gives*, *results in*, any one of which is itself multi-meaning. There may well be other conceptual factors involved, but this linguistic diversity seems likely to be implicated in the findings that children experience difficulties in interpreting the equals sign well into their school years (Baroody and Ginsburg, 1983; Cobb, 1987).

The reader can determine swiftly for most of the words listed in Table 7.1 what meaning each is likely to have for the child before he or she encounters it in a mathematical context at school. Not every word will be acquired before school entry, of course, but in most cases it is reasonable to assume that the child will first learn the word in a non-mathematical context.

Although we can step back from this vocabulary list and spot ambiguities, such metalinguistic reflection is not necessarily available or practicable in the classroom. Hence, the question arises of whether these ambiguities cause any problems for pupils. After all, many words are ambiguous, but in everyday speech we take advantage of situational and linguistic cues to determine the

Table 7.1 Some ambiguous words used commonly in school mathematics

above, altogether, angle, as great as, average, base, below, between, big, bottom, change, circular, collection, common, complete, coordinates, degree, difference, different, differentiation, divide, down, element, even, expand, face, figure, form, grid, high, improper, integration, leaves, left, little, low, make, match, mean, model, moment, natural, odd, one, operation, overall, parallel, path, place, point, power, product, proper, property, radical, rational, real, record, reflection, relation, remainder, right, root, row, same, sign, significance, similar, small, square, table, tangent, times, top, union, unit, up, value, volume, vulgar.

contextually appropriate sense. If a friend says 'I am going to withdraw my savings from the bank', we are unlikely to infer that she hoards her wealth somewhere by the river. Do maths pupils exploit context in a similar way to recognize that a *mathematical operation* does not invariably require the administration of anaesthetic and intrusive surgery, that *rational numbers* are not necessarily more dependable than the odd ones, or that a *vulgar fraction* is not a little boy who swears a lot?

Not a great deal of research has been conducted into these matters, but a tentative answer is 'no, not always'. There are many anecdotal reports in the maths education literature of pupils misinterpreting maths expressions, instructions or questions in ways which show clearly that they are attempting to decode an ambiguous word in its everyday sense. For example, Pimm (1987: 8) mentions a 9-year-old who responded to the question 'What's the difference between 24 and 9?' with the (irrefutable but contextually inappropriate) observation 'One has two numbers in it and the other has one' (and see Cowan, this volume, on the interpretation of *same* and *different*). Hart (1981: 21) reports the following exchange between a secondary school pupil and an interviewer who was probing on maths concepts:

Interviewer: Do you know what volume means?
Child: Yes
Interviewer: Could you explain to me what it means?
Child: Yes, it's what is on the knob on the television set.

A charming example is provided by Hobart (1980) of a child finding difficulties with questions that began 'What is common to . . .?', because of the previously acquired understanding that *common* meant 'too much lipstick'.

Practising teachers offer many such examples. This suggests that difficulties with lexical ambiguity are experienced widely, and indeed it is conceivable that not all come to the attention of teachers. Nicholson (1977) reported that only about 20% of a sample of secondary pupils taking the British CSE understood the word *product* as used in mathematics. Teachers may use multi-meaning terms in ways which their pupils misunderstand, and pupils working independently may impose incorrect interpretations on maths vocabulary. One context in which this may occur where teacher-feedback cannot always be immediately available is in textbook and workcard activities: if children are set exercises based on written materials (a practice which of course has increased with the advent of the new maths), their understanding of the instructions/tasks is crucial, and ambiguous words could lead to confusions.

We tested this possibility experimentally in a reading task administered to 7- to 10-year-old pupils (Durkin and Shire, in press). The task (based on work by Mason *et al.*, 1979) required children to read sentences which contained ambiguous words of the kind we have been discussing: words such as *make*, *table*, *times*, *big*, *above*, etc. Each word appeared in two sentences. One of the sentences used the word in its everyday sense, e.g. 'Mum and John *make* cakes.' The other sentence used the word in a mathematical sense, e.g. 'Two and two *make* four.' The child's task was to identify, from a set of options presented below the sentence, what the target word meant (the target word, in this case *make*, was underlined). The sentences were presented in booklets, administered on two separate occasions, and arranged so that the two sentences for any one target word appeared in separate booklets.

The response options included one word which was a near synonym of the everyday sense of the target (in this case, *cook*), and one word which was a near synonym of its mathematical sense (in this case, *equals*). In addition, there was one word which was related thematically to the everyday sense (*eat*) and one word which was related thematically to the mathematical sense (*add*). These options were presented in random order. A fifth option, always presented last, was 'No answer'. Children had to tick one box per sentence to indicate what they thought the target word meant in that sentence.

Mason *et al.* (1979) had found that children were prone to an error pattern in which they would identify the dominant sense of an ambiguous word even when the sentence context was biased toward the subordinate sense. We found very similar results with our sentences: when children misidentified the meaning of an ambiguous word in a mathematical sentence, the sense they chose was often the everyday sense. This happened significantly more often than the reverse type of error (i.e. interpreting an everyday use of an ambiguous word as though it conveyed its mathematical sense).

These errors did not occur on every trial and older children in the primary age range we tested produced less errors than did younger children. Nevertheless, the fact that systematic error biases are detectable toward the everyday meanings of the words suggests the possibility that at least some pupils sometimes make such misinterpretations in the course of their use of written materials in mathematics, and errors were still occurring even among the older children in our sample. Lexical ambiguity among words that adults take for granted may occasion serious misunderstandings for pupils.

Spatial terms in mathematical descriptions

One particular type of lexical ambiguity that is pervasive in mathematical description concerns the use of spatial words – *high*, *low*, *up*, *down*, *top*, *bottom*,

1 2 3 4 5 6 7 8 9

Fig. 7.1. The conventional number line. Which is a high number?

○ ○ ○ ○ ○ ○ ○ ○

Fig. 7.2. Which is a high circle?

big, *little*, etc. These words are very frequent in everyday speech, and children begin to learn their basic (spatial) meanings early in language acquisition (Furrow *et al.*, 1985–6; Tomasello, 1987). By the time children begin formal education, they can produce spontaneously and appropriately many of the common spatial terms and show good comprehension ability for all but the most complex (such as *in front of/behind*, *between*: see Durkin, 1981). However, when they begin school, children are likely to hear spatial words used occasionally to describe relations among numbers, e.g. in utterances such as 'the numbers are going up', 'eight is higher than five', 'the bottom number is one', and so on.

The potential for ambiguity here is considerable. For example, consider the familiar array of numbers in Fig. 7.1. This is a standard way of presenting written numbers in Western education, and most elementary maths curricula would aim to ensure that children become familiar with the number line early in their maths experiences. However, note that when we say, for instance 'eight is higher than five', we mean to convey a relationship rather different from that intended when we say 'the kite is higher than the tree top'. In the same way, when we say 'the numbers are rising', or 'show/tell me the numbers below four', we (adults) are working normally on a numerical frame of reference rather than a literal, spatial one.

If a teacher presented a child with an array as set out in Fig. 7.2 and requested 'Show me the low circle', we would reject the task as perverse and unlikely to reveal much about the child's linguistic competence.

Spatial words are used polysemously in numerical descriptions (Durkin *et al.*, 1986a). The basic meanings, concerning locations in physical space, are extended into a new domain where the critical dimension is relative numerosity. Once you know that, you can follow the kinds of simple number descriptions given above. But how do you learn the new use of old words, and how do you determine when one sense is intended rather than another?

To make matters still more complicated, we use spatial words when describing arithmetical operations, such as *adding up* (where the numbers are arranged vertically and the answer is required at the bottom). Even in everyday use, these minor words can cause occasional confusion: Hutt (1986: 1) describes a language-disordered 7-year-old's insistence that a person cannot be *cleaning up* if she is (down) in the cellar. Barham and Bishop (this volume) note the difficulty for teachers of deaf children of *signing* phrases such as *add up*, *left over*.

As in much of language acquisition, these subtleties are rarely communicated

Fig. 7.3. A number tree: pick a high number. Drawing by Linda Jeffery.

explicitly. Teachers do not provide infant school children with lessons in metalinguistics and polysemy – at least, not intentionally. Teachers are likely to vary in the extent to which they are aware of this area of potential problems; in the course of many discussions with teachers' groups, we have found that some have experienced and tried to alleviate problems in this respect, while others have not considered the issue before. However, all language users are likely to use the

Ring each set of large animals

Ring each set of smaller animals

A set of large animals and a set of smaller animals

Fig. 7.4a. Sets of animals intended to reinforce children's ideas about large and small objects (a modified version of the ideas presented in Howell *et al.*, 1980). Drawings by Linda Jeffery.

words we are talking about in their everyday senses even in maths lessons, because they are high-frequency words for which we have many uses. Hence, the input to children is likely to include examples of the terms used in their everyday senses *and* in their specialized, mathematical senses.

Maths texts, too, provide illustrations which are themselves ambiguous. For example, Fig. 7.3 shows a number tree similar to illustrations found in early number books. If you were asked to pick a high number from the tree, what would you opt for? Fig. 7.4 is based on examples found in Fletcher maths, Level 1

Fig. 7.4b. A game intended to strengthen knowledge of numerical sequence (a modified version of the ideas presented in Howell *et al.*, 1980). Drawing by Linda Jeffery.

(Howell *et al.*, 1980). Figure 7.4a is intended to consolidate children's understanding of *large* and *small*. But are the 'smaller' animals small, or distant? Are mice and rabbits large when you have just been thinking about kangaroos and sheep? Figure 7.4b provides experience of progress *down*(?) the number sequence ending *up*(?) at a *high* number.

Similarly, Hanley (1978: 28) points out that primary texts in mathematics often introduce children to the words *vertical* and *horizontal* with the advice that 'Vertical lines are drawn up and down, Horizontal lines are drawn across' (cf. Griffiths, 1969). The problem arises that this presupposes that the reader allows for the conventions of representation of real space on paper: strictly speaking, a line drawn in any direction on a page resting on a normal table surface is horizontal (and see Hanley, 1978, for further discussion with reference to the notion of upright line graphs). Correspondingly, for you as a literate reader the top of this page is up where the running head appears – but if top means 'uppermost surface' an alternative interpretation is possible.

These observations lead to the question of whether children are inclined to interpret the spatial words we have been discussing as spatial even in mathematical contexts. There is evidence that young children do this quite often, and that even older children do so sometimes. We summarize some of this evidence in the next section.

Children's understanding of spatial terms in mathematics

We have conducted a number of studies of young children's interpretations of spatial terms applied to numbers. In a preliminary study, we asked 4- to 5-year-olds simply to write for us *high* or *low* numbers. Figure 7.5 presents examples of typical responses.

Adda (1982) describes a similar case in which a 9-year-old was asked by his teacher to write a big number on the blackboard, and he provided in very large numerals 352. In some instances, children are indifferent to the number they actually draw and the critical dimension for them appears to be the spatial one; in others, children write a numerically higher number in the higher spatial location, and a lesser number in the lower location, thus achieving a congruence of spatial and numerical senses.

Unfortunately, as we have indicated above, some materials that are available to children may not afford this congruence, and the possible interpretations of the key words may conflict. Such instances are useful from the researcher's point of view because they allow for a test of which sense has priority for the child – spatial or numerical. To conduct such a test, we presented children with an upright model of a house which had four windows (two up, two down) into which we could insert numbers (on a card which slipped behind the façade of the house). We presented two numbers at a time, such that the numbers could be vertically or horizontally arranged, and they could be identical or different in terms of numerosity represented. Thus, spatial and numerical senses were sometimes congruous (e.g. a number 8 was presented above a number 4), sometimes conflicting (number 3 above number 7), sometimes only the numerical sense gave a basis for a response (number 4 and number 8 presented side-by-side) and sometimes only the spatial sense gave a basis for a response (two number 7s, one in a top window, and one in the window below it). The critical condition is the conflict one, and we found that 4-year-olds tend to opt for a spatial strategy here, whereas 5-year-olds are beginning to employ a numerical one. In a modified version of the experiment with slightly older children, we found that biasing the context towards either space or number affected whether children opted for spatial or numerical responses.

Durkin *et al.* (submitted) tested children aged 3 to 8 years for their understanding of the term 'big number'. Again, using a picture-choice task, the

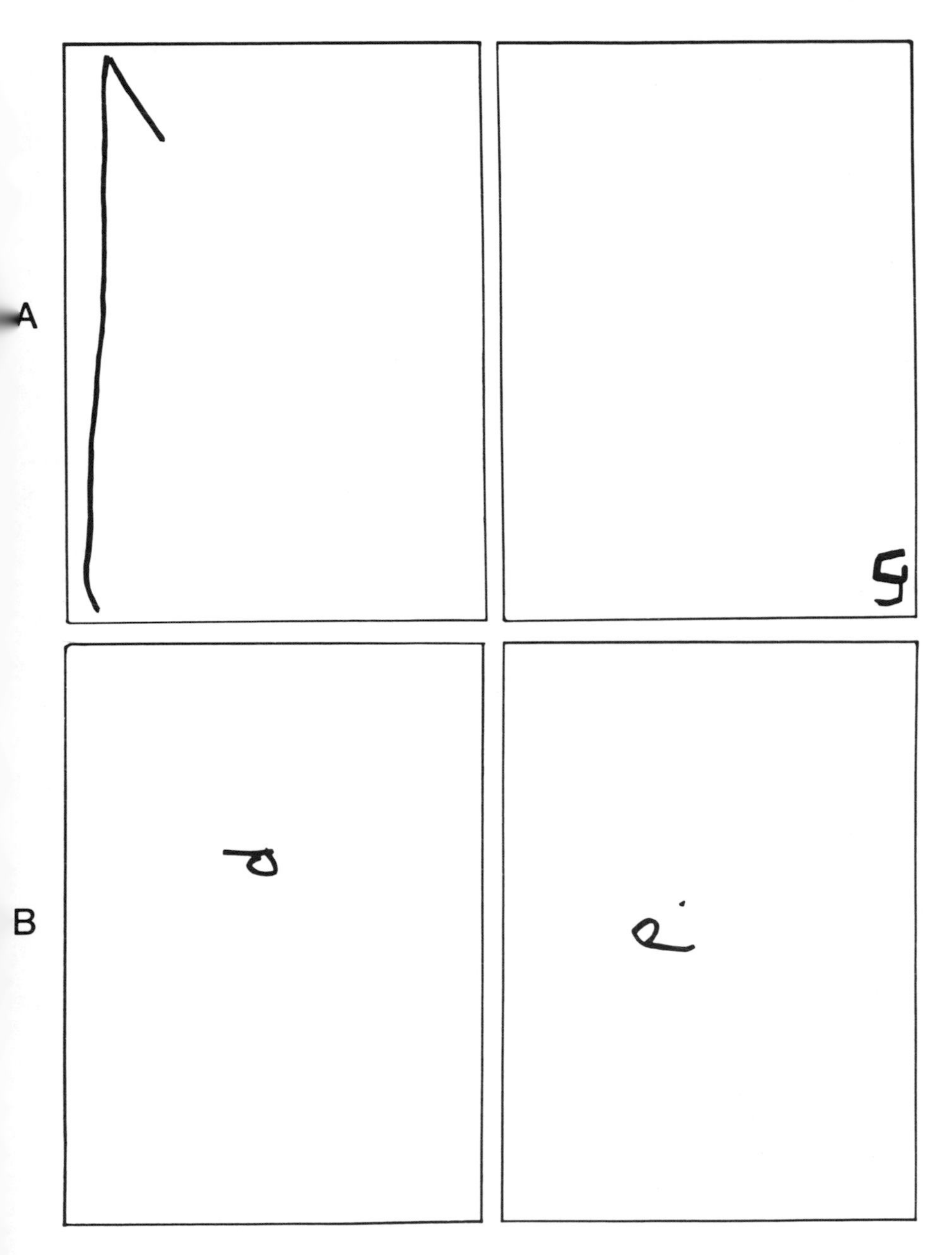

Fig. 7.5. Young children's written productions of 'high' and 'low' numbers. Each child's reponse to a request to write a 'high number' is shown on the left and a request to write a 'low number' is shown on the right. Reproduced from original productions on A4 paper. Child A (aged 4 years) said his numbers were 'one' and 'five', respectively. Child B (aged 5 years) said her numbers were 'six' and 'nine' respectively.

basis for selection could be numerical status or physical size. In cases of ambiguity, children throughout this age range opted predominantly for the size criterion.

Although these responses are prevalent among younger school children and are probably resolved for the majority during middle childhood, it should be noted that even secondary school pupils experience difficulties with some spatial terms applied to number relations. For example, Bell (1984) reported that children of this age produce many errors in response to instructions to rank items moving up and down a table or ladder. Given 'Tottenham Hotspur was in 8th place and has moved up 3 places, where is it now?', many pupils responded '11th'. We replicated these findings with upper primary pupils (Shire and Durkin, 1989).

In short, children do appear to manifest some uncertainty with the very basic terms of mathematical descriptions that we have been discussing. Importantly, their responses indicate a systematic bias in interpretation, and not surprisingly this is to strive to interpret the term in its basic, spatial sense. We will not discuss the psycholinguistic issues in detail here (see Durkin *et al.*, 1986a, for a fuller account), but it should be clear that children appear to exploit the linguistic information that they have been acquiring and using in everyday contexts prior to, and independent of, these school experiences. Again, it is easy to imagine actual situations in the classroom where such a bias could be disadvantageous to the pupil. Indeed, if people seem to be using familiar words to mean new relationships, it is easy to see how mathematics could become a distressing experience.

Practical implications

The issues we have been discussing raise direct concerns for educators, and we suggest here several practical strategies that can be adopted.

1 *Monitor lexical ambiguity.* In preparing materials for use in maths teaching (from lesson plans to textbooks), it is advisable to consider whether any words might have a different meaning for the pupil from that intended or assumed in this specialist context. Table 7.1 provides an extensive list of possibilities. If the teacher is aware in advance of potential problems, this increases his or her sensitivity to pupil performance.

2 *Enrich contextual cues.* Successful decoding of a specific use of an ambiguous word is dependent upon use of context. The contexts provided by maths lessons may not always be sufficiently familiar or secure for the pupil that he or she can be expected readily to access the appropriate mode of discourse and terminology. For the proficient, this may not be a problem, but for children who are finding maths a difficult area, confusions over the specialist use of terminology can only add to their discomfort. Creative and sensitive preparation can help guide the child to the context-specific meaning of the target words. Shire and Durkin (1989) found that providing pupils with

visual support for judgements about rank ordering facilitated their performance.

3 *Exploit ambiguity to advantage.* In many cases, particularly with polysemous words, purposeful use of those contexts in which the everyday and specialist meanings of the terms coincide can provide pupils with a secure base in which to familiarize themselves with the new extensions of their vocabulary before venturing into more demanding uses. For example, we have developed video-based lessons and practical games that teach infant school children about the uses of spatial words in number contexts by displaying, talking about and working on numbers in spatial locations congruous with their numerical status (e.g. arranging illustrations so that a numerically high number will always be spatially higher than numbers representing lesser numerosity). In training experiments, we found that children's understanding of the numerical uses of the key terms can be enhanced in this way.
4 *Confront ambiguity.* Although we have emphasized the liabilities due to lexical ambiguity in this chapter, we do not wish to convey that multiple meaning is a pernicious conspiracy designed to thwart the advancement of education. On the contrary, there are good reasons, in terms of linguistic economy and organization, why homonymy and polysemy exist (cf. Durkin *et al.*, 1986b; and see Nolder, this volume, on the advantages of metaphorical extensions). It may be, though, that children's appreciation of the varieties of meanings can be promoted by developing teaching materials which bring potential conflicts into focus. Again, this calls for careful monitoring of pupils' responses, but in our work we have found that children can profit from such encounters. This is likely to be particularly so from middle childhood, the period during which children become more sensitive to linguistic ambiguities in general (a period indicated by increasing interest in word play, jokes, puns and riddles: see Shultz and Horibe, 1974).

Conclusions

As we stressed at the beginning of this chapter, we have been concerned here with only one aspect of linguistic ambiguity in mathematics education and we would not propose that all of children's difficulties in mathematics can be traced to misinterpretations of multi-meaning words. Indeed, we do not suppose that every encounter with an homonymous or polysemous term in maths will result in the child interpreting the term in its basic sense; clearly, children would never progress in the subject if this were the case. It may be that some potential difficulties with lexical ambiguity are experienced only fleetingly during early exposures, though this in itself would merit further attention as a feature of the teaching context. No doubt there are individual differences in the extent to which children are able to resolve the linguistic difficulties we have been considering, with or without assistance; though again, we would expect that those faring less well in maths classes will be most vulnerable to problems. The evidence from several sources reviewed above suggests that for at least some pupils, semantically based misunderstandings do arise through the primary school years, both in oral exchanges and in written work. Many of these difficulties can be identified and anticipated by attention to the possible meanings of key vocabulary, and this enables us not only to understand specific problems but to develop appropriate teaching strategies.

Acknowledgements

The research described in this chapter was supported by grants to Kevin Durkin from The Leverhulme Trust, UK, and the Special Research Fund, University of Western Australia. We are grateful to Linda Jeffery for artwork and to Herb Jurkiewicz for the photographs.

Further reading

Durkin, K. (ed.) (1986). *Language development in the school years*. London: Croom Helm (esp. ch. 5).

8 The language of fractions

Daphne Kerslake

The manipulation of fractions has often been used as a measure of success at mathematics and has also traditionally generated anxiety in the hearts of learners. Despite the fact that there is little reason why the majority of young people should indeed be concerned with other than the recognition of fractions and simple comparisons, particularly since the ready availability of calculators, concern is still expressed about the learning of fractions. C. Frusher Howard, in his *Art of reckoning*, published in 1887, observed: 'the system of decimal fractions is so eminently simple that when it is generally understood will entirely displace the clumsy system of common fractions'. That time does not yet seem to have arrived. Indeed, in the UK, a member of the Mathematics Working Group on the National Curriculum and Attainment Targets ignored the sensible comments of the majority and broadcast his regrets that multiplication of fractions was not to be included in the curriculum for all children under the age of 11 years. Yet, were he to be challenged to find instances in which one needs to multiply or divide fractions in everyday life, he would find it impossible to find any but very contrived situations. The few genuine instances of multiplication, such as that used by John Greig in his *The young ladies guide to arithmetic* (1858) – 'If a yard of ribbon cost 5d., what will 16 yards cost?' – have long since gone because of the introduction of metric units.

This chapter discusses some of the language issues in the field of fractions and, in particular, examines the use of phrases such as 'cancel', 'top heavy', 'of means multiply', 'share by', 'divide into', 'multiply means make larger, divide means make smaller'. It will make reference to some research findings on children's understanding of fractions; consider some of the difficulties encountered in the learning of fractions; discuss several models of a fraction and their effect on the learner; the use of inappropriate and limiting language and the role of algorithms; and inappropriate generalizations.

Description of research on children's understanding of fractions

The research on fractions which formed part of the Strategies and Errors Project based at Chelsea College, London, 1980–83, chose to concentrate on the basic understanding of the concept of fractions, their comparison and the notion of equivalence. The children who took part in the study were in middle-ability classes and were aged 12–14, and it was clear that their problems with fractions could be attributed to the fact that their perception of a fraction was very restricted. With their only model being that of a part-of-a-whole, and this usually part of a circle, they found it virtually impossible to think of a fraction as a number. Consequently, they could not attach any meaning to the notion of addition, let alone multiplication or division.

Although the difficulties, then, were due to an inappropriate model, there were also undoubtedly certain interesting language issues which emerged during a period of extensive interviews with individual pupils.

Fractions as objects

It can be argued that many of the difficulties which pupils experience are due to the restricted and imprecise use of language in describing fractions. The most common apparatus used in primary classrooms for illustrating fractions consists of a single shape, square or circle, and a set of smaller pieces which together make up the whole. Typically, a semicircle will be labelled 'one half' or $\frac{1}{2}$; yet these are whole pieces in their own right. Different shapes for different sizes are all supposed to represent the same fraction! Such apparatus requires the learner not to see individual pieces as identifiable shapes with their own identity, but rather to regard them as parts of some other complete shape.

Therefore, from an early age, children associate fractions with concrete things. This is analogous to the way in which counting numbers are presented: by the focusing of attention on 3 books, 3 children, 3 pencils or 3 anything, children begin to make a generalization of the abstract concept of three (see the chapters by Fuson and Sinclair, this volume). Teachers are aware that such exemplars are but a stage in the acquisition of number. But it seems that, for many children, we do not appear to replicate this process in the case of fractions. For many people, fractions are seen only in terms of their concrete representations.

Mary Everest Boole, writing in 1931, wrote:

> It may occur to some that too much use is made of examples relating to food. But we cannot make mathematicians by insisting upon a non-existing superiority to physical facts. Apple or bun forms the natural unit for the child; the sharing of a cake or a fruit is the natural fraction as well as the true introduction to the ethical life.

Ethical considerations apart, what Mary Boole was saying is all right as far as it goes, but experience suggests that, for many, our discussion about fractions stays at the level of parts of apples or buns. Yet, at the same time, children are expected to understand and use operations such as addition and multiplication of fractions. There is no possible meaning that can be attached to adding one-half of an apple to, say, two-thirds of a bun: it is numbers that are added; not pieces of food or geometric shapes. Multiplication is even more inexplicable: try to interpret the product of $\frac{2}{3}$ of a bun and $\frac{3}{4}$ of an apple! We need to discuss these issues with the learner, and to find ways in which this new view of fractions as numbers can be accommodated.

Children's models of fractions

It will be predicted from the above, that when children were asked to say what they thought a fraction was, the majority referred to pieces of cake or some other part of a whole. Typically, they said 'Fractions are like pieces of cake being cut into different pieces.' Of those who appeared to recognize some connection with numbers, some interesting comments were made. A few reflected the Latin derivation of the word in defining fractions as 'broken up numbers' or 'I think fractions are numbers broke into bits.' One can only guess whether they think they remain as numbers after being broken. Several said that fractions have to be less than 1 (the part–whole problem again). Other connections between fractions and numbers were expressed:

> Fractions are the in-betweens of numbers, not quite whole numbers.
>
> I think fractions are spilt [*sic*] from whole numbers.
>
> They are whole numbers split into smaller numbers.
>
> They are unhole [*sic*] numbers.

and the ambiguous:

> A fraction is not allways [*sic*] less than a numer [*sic*].

One wonders how often, if at all, pupils have actually heard any teacher say that fractions are numbers. At some stage in our education process, our view of fractions as parts of geometric shapes – the ubiquitous introduction to fractions – has altered as we have moved towards the idea of rational numbers. Yet experience suggests that this transition is rarely made explicit or discussed, nor are we aware that it has taken place.

A second concern relates to the link between the fraction $\frac{3}{4}$ and the expression $3 \div 4$. During the interviews, many pupils denied that there was any connection. When asked to interpret the expressions $3 \div 4$, $4 \div 3$, $12 \div 4$ and $4 \div 12$, a variety of linguistic forms were used by the pupils. Many used the expression 'shared by', as in:

> It's 3 shared by 4 . . . that's 3s into 4 . . . 1 remainder 1;

or:

3s into 4. 3s go into 4 one, and one remainder.

3 shared by 4 . . . 3s into 4, isn't it?

Much has been, and could be, said about the way children interchange the order of the division. Indeed, many of those interviewed clearly believed that $3 \div 4$ and $4 \div 3$ were two expressions for the same thing. One child, asked about $4 \div 12$, said: 'That's the wrong way round . . . 12 won't go into 4. Change it round – 12 shared by 4.' When asked if it was all right to choose which way to do it he said: 'Yes, because you'd get that one otherwise – 12 into 4 and that's the wrong way.'

The language of sharing and division

For the moment, it is worth stopping to consider the use made by children of the verb 'to share by'. It is a curious yet very common expression, and it is difficult to see either what the user thinks it means or where it comes from. The notion of sharing items between children is probably everyone's first introduction to the idea of division. That is a natural form of speech – you have things such as sweets and children who all want a fair share, so you share them equally between them. So what does 'to share by' mean? The notion of sharing sweets by children is bizarre! The confusion lies in the comparison with 'dividing by', which applies strictly to numbers only. Division is a generalization of two different physical activities. One aspect is, for example, that of sharing a certain number of items between a number of people. The other involves the idea of grouping, as when one finds how many groups of 4, say, can be made out of 12 items. We have two quite different physical activities, with two different results, but which are both represented by the same symbolic form.

An alternative form used by several children was 'you divide the 4 into the 12' or 'you find out how many times 4 goes into 12'. The notion of 'going into' is worth considering. What goes into what? What exactly do we mean when we say '4 goes into 12 three times'? What image do the words present? Do they help us to appreciate that 12 divided by 4 is 3 or that three 4's are 12? And what is the difference between 'dividing by' and 'dividing into'?

Classroom jargon

The world of the classroom is cluttered with many meaningless phrases, ones we all use without as much as a thought. 'Sharing by' and 'dividing into' are but two of many. It is difficult to determine to what extent such phrases affect the process of learning. What is certain is that children pick up and use the phrases and, we assume, that they absorb the ideas at the same time. We do not know what meaning they have for the child, nor what image they present. At best, they carry with them some general view of a sharing or dividing activity. At worst, the child learns that word and ideas do not necessarily have any connection. We do not know how many children – if any – have experienced difficulties that can be attributed to our classroom jargon. It has to be perverse, though, to use phrases which do not reflect the processes being used. Algorithms, such as those for finding equivalent fractions or for adding or multiplying fractions, are rich in such ill-defined and unhelpful phrases. Some get recited almost as if they were part of a jingle, and show evidence of having been rehearsed many times. Some examples of this are now given.

Children's language in applying learned algorithms

During the course of interviews, pupils were asked how they would deal with equivalent fractions and the addition of fractions. L.G. was asked to work out $\frac{2}{3}+\frac{3}{4}$, and this was part of her response:

$$\frac{3}{2}+\frac{3}{4}$$

L.G.: You've got to make them the same. I think you turn one over. Yes. Oh, but that's not right . . . they've got to be the same at the bottom. It's made it top-heavy. Um . . . you could cancel down first . . . two goes into itself once, and two goes into four twice . . . three goes into three once, so that makes that $\frac{2}{3}$.

Interviewer: What do you think about that?

L.G.: Yes.

'Turning over' and 'cancelling' are very frequently tried as recipes for success in dealing with fractions. G.L. was interviewed about some equivalent fractions and, in the following extract, was trying to find the value of the missing number in the expressions $\frac{5}{3}=\frac{15}{?}$ and $\frac{9}{12}=\frac{12}{?}$.

G.L.: $\frac{5}{3}=\frac{15}{?}$ . . . You times that [the 15] by three, so you times the three by three, which is nine . . . $\frac{15}{9}$

Interviewer: Fine. What about this one: $\frac{9}{12}=\frac{12}{?}$?

G.L.: Ooh, crumbs! Three goes into nine goes three, and three into 12 goes four . . . so 12 into . . . it would be nine. $\frac{9}{12}=\frac{12}{9}$. That's top-heavy . . .

Interviewer: And was the $\frac{9}{12}$ top-heavy?

G.L.: No. . . . Um, you could divide it again . . . you could cancel it down. Three into 12 goes four, and three into nine goes three. $\frac{12}{9}=\frac{4}{3}$. That's still top-heavy. You could cancel $\frac{9}{12}$ down.

Interviewer: Does that help?

G.L.: It can't be right. It could be $\frac{12}{3}$. That's four whole ones.

Interviewer: Is that all right? $\frac{9}{12}$ equals four whole ones?

G.L.: No! It could be $\frac{12}{12}$ . . . 12 goes into 12 once . . . oh no! . . . $\frac{12}{4}$. That looks all right.

Here, G.L. was confident over the first equivalence, but was disturbed by the existence of a 12 on both sides of $\frac{9}{12}=\frac{12}{?}$. The solution of the problem involves multiplying by a fraction, namely $\frac{12}{9}$ or $\frac{4}{3}$ or of increasing 12 in the ratio 12:9. This proved very hard. Indeed, G.L. was one of the very few interviewees who even tackled the task. He recognized that his attempt at $\frac{12}{9}$ gave him a 'top heavy' fraction – and so he appeals to cancelling, in the hope that it would help. Having found it to be $\frac{4}{3}$, he realized that he had not solved the problem of 'top-heaviness', but he made the very sensible suggestion that the original $\frac{9}{12}$ could be cancelled. This would, in fact, have made the question much easier, but, unfortunately, he did not pursue it and returned to his $\frac{12}{9}$ with no success.

Cancelling, or rather the attempt to cancel, was probably the most ill-used strategy observed during the interviews. Two typical absurdities that injudicious cancelling led to were:

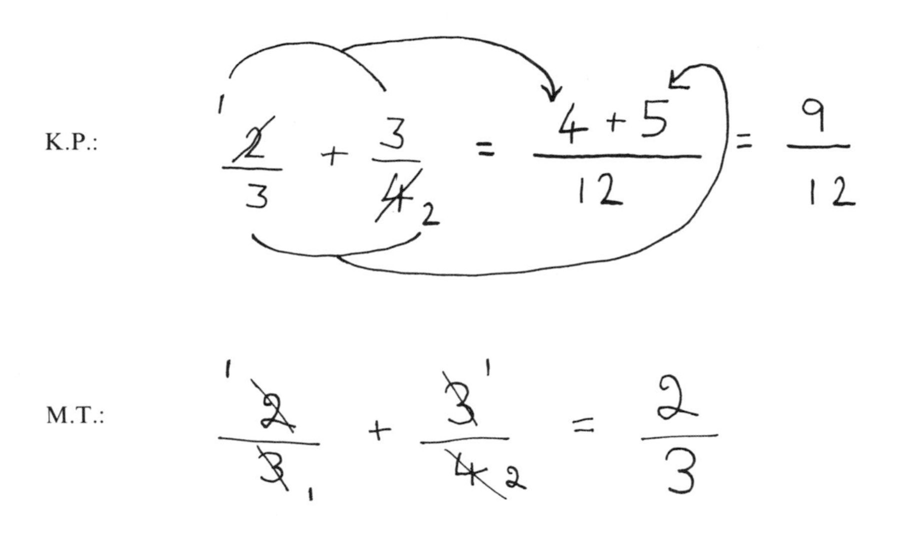

> . . . that's $\frac{2}{3}$. I don't think that's right . . . you could times them . . . $\frac{17}{17}$ . . . don't think that's right either. . . . Take them away? You could say 2 from 4 – no, that doesn't sound right.

One might question whether the use of the word 'cancel' is another instance where the word gets in the way of the concept. The use of the word 'to cancel' in general speech has the idea of to annul, to remove, to undo, to countermand or to neutralize. This may explain the way many children appealed to the notion of cancelling to get them out of a difficulty. But this does not reflect what actually happens when a fraction is 'cancelled'; it is yet another example of the way we use a word in general use, attribute a specific mathematical meaning to it, but never refer to what we have done!

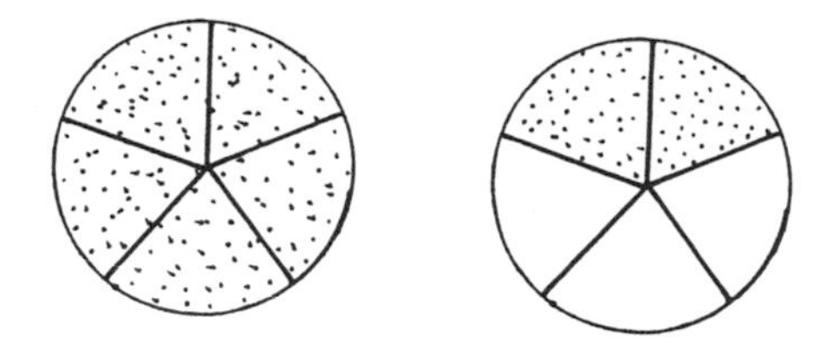

Fig. 8.1. Circles as stimuli for fractions tasks.

Both L.G. and G.L. were concerned about their fractions being 'top-heavy', a phrase that occurred very frequently; though, curiously enough, the alternative 'bottom-heavy' rarely arose. Although the idea of comparing the size of the numerator and the denominator seemed to be well understood, one might again question the origin of the use of the word 'heavy'. One wonders how the concept of weight crept into the vocabulary of fractions! Possibly it was the notion of a fraction being either stable or about to topple over – but it is a curious phrase to use when what we really mean is to distinguish between fractions which are either greater than or less than one. The focusing of attention on this particular criterion also reflects our preoccupation with the part-of-a-whole model of a fraction, which really limits us to fractions less than 1. Attempts to persuade interviewees to interpret Fig. 8.1 as a fraction greater than 1 were not very successful. When asked 'If you took all the shaded pieces from these circles, how much would you have?', $\frac{7}{5}$ and $\frac{7}{10}$ were proferred with about the same frequency. Similarly, when shown Fig. 8.2, there was confusion as to whether it was $\frac{9}{8}$ or $\frac{9}{16}$.

'Making the bottoms the same' was the frequently quoted recipe for adding fractions with different denominators. The reason for so doing was usually given as 'You can't add them unless you do.' For example:

Interviewer: Why did you use the 12?
S.E.: Because these two – $\frac{2}{3}$ and $\frac{3}{4}$ – don't add up together. So you've got to make the bottoms the same – find a number what they both go into.
Interviewer: Would you explain why that does make it work?
S.E.: Because a 12 and a 12 are the same, and a third and a fourth aren't. So I timesed the three by four to get 12.
Interviewer: OK. Now, someone might say that you've done a different question?'
S.E.: Yes, but you wouldn't be able to do that properly $\frac{2}{3}+\frac{3}{4}$ . . . without doing complicated sums and everything, but with that, it's a lot easier.

There was a sense in the responses of many of the pupils in the interviews that the original fractions had not been added at all – that finding a common denominator was a way of changing the task so that it turned into a pair of fractions that could be added, i.e. that the original fractions had been replaced by more convenient ones. Consider these two extracts from interviews with P.E. and G.W., in which they were both talking about adding $\frac{2}{3}$ and $\frac{3}{4}$:

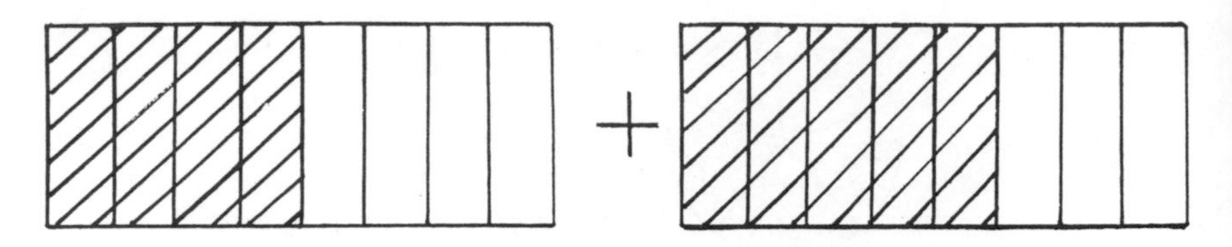

Fig. 8.2. Rectangles as stimuli for fractions tasks.

Interviewer:	Could you explain why you did that?
P.E.:	To make the bottoms the same fraction.
Interviewer:	And why do you have to make the bottoms the same?
P.E.:	Because you can't add them if they are thirds and fourths.
G.W.:	You have to find the number the three and the four will both go into. And then how many times three will go into 12, four, and then four times two is eight . . . that's $\frac{17}{12}$ . . .
Interviewer:	Fine. Instead of writing $\frac{2}{3}+\frac{3}{4}$, you wrote $\frac{8}{12}+\frac{9}{12}$. Somebody could think that was a different sum you're doing?
G.W.:	Yes.
Interviewer:	What would you say to them?
G.W.:	You have to put them in twelfths if they're not the same.
Interviewer:	But someone might think you're doing a different sum?
G.W.:	Yes . . . it's just the way we've been taught.

'Just the way we've been taught' describes the unquestioning acceptance of what teachers say.

'Of' means 'multiply'

One fact that appeared to be fairly universally appreciated was that, when encountered in the context of fractions such as 'Find $\frac{2}{3}$ of $\frac{3}{4}$, the word of means multiply. Why is this rule seen to be so necessary, and why is it true? The rule has been with us for a long time. In *The instructor – or young man's best company*, George Fisher (1860) wrote: 'The word "of" placed between two fractions means exactly the same as the sign of Multiplication.' Presumably, the rule derives from the first introduction to multiplication for most young children – that of evaluating, for example, four *lots of* three. The 'of' survives through, say, $\frac{2}{3}$ of a *lot of* 12, to $\frac{2}{3}$ of a *lot of* $\frac{3}{4}$ – by which time any reasonable interrpretation of a 'lot of' anything has disappeared. Without any discussion of the matter, of course! The fact that when '3 lots of 4' gets recorded as 3×4, it is then referred to as '3 multiplied by 4' – an altogether different interpretation – is yet another hazard.

Fisher does, unusually, actually discuss the need to adapt one's view of multiplication in order to apply it to fractions. He writes: 'You are to observe that Multiplication in Fractions lessens the product, though in whole numbers it augments.'

'Multiply makes bigger' and other inappropriate generalizations

This comment of Fisher's leads to the final point of this chapter, i.e. that children are given, in their early years, global 'truths' such as 'multiply means make bigger', 'divide means make smaller' and 'you always divide the smaller number into the larger number'. These are true enough when working with whole numbers, or counting numbers, only. But they certainly do not still apply when the range of numbers being used is extended to include fractions. Yet, once again, it seems that pupils are expected to accommodate to a new definition of both multiplication and division. What is more, we find that it is possible for two numbers – if they are fractions – to be 'the same'; something that was certainly not true of a number like '3'.

Mathematics is hard for many people to learn; we do not make it any easier by using ill-defined words and by changing the interpretation of others without even a minimum acknowledgement.

Practical implications for teachers and textbook writers

1 Although the part–whole model of a fraction has its use, much more attention needs to be given to other models such as 'three items shared between 4 people'.
2 The fact that fractions are numbers needs to be emphasized. More use should be made of number-lines, with fractions marked as points on the line.
3 Calculators can be used to reinforce the idea that $\frac{3}{4}$ is $3 \div 4$.
4 The use of words such as 'cancel' and 'of' need to be discussed and understood.
5 The language of sharing and dividing should also be discussed.

Conclusions

Fractions are commonly regarded as difficult to learn and to teach. Although the use of calculators has, inevitably, resulted in the study of fractions being seen as of decreasing importance, it is impossible for them to be totally disregarded. Some of the difficulties can be attributed to:

1 A view of fractions that is limited to a part–whole model.
2 The existence of some inappropriate models of fractions (such as 'this is half a square').
3 A reluctance to accept that fractions are numbers.
4 A dependence on algorithms that have been learned by rote.
5 Confusing terminology such as 'share by' and 'divide into'.
6 A failure to recognize that properties that were true within the field of counting numbers are not necessarily true for fractions.
7 The emphasis on partial truths at earlier stages which no longer hold in the case of fractions (such as 'multiply means make bigger').

Further reading

Booth, L.R. (1981). Child methods in secondary mathematics. *Educational Studies in Mathematics*, **12**, 29–41.

Hasemann, K. (1981). On difficulties with fractions. *Educational Studies in Mathematics*, **12**, 71–87.

9 The language of multiplication and division

Julie Anghileri

In most traditional school texts, the four operations of arithmetic are introduced in the sequence: addition, subtraction, multiplication, division. Although they are introduced as distinct operations, their separation in the early stages of learning may inhibit children's understanding of the fundamental links between addition and subtraction and between multiplication and division, and the fact that much of the language for each pair is shared. Indeed, in higher mathematics, subtraction and division cease to have independent identities as the ideas of inverse operations and inverse elements are developed. At this level, $6-3$ is associated with $6+(-3)$, (-3) being the *additive inverse* of $(+3)$, and $6\div 3$ is associated with $6\times\frac{1}{3}$, $\frac{1}{3}$ being the *multiplicative inverse* of 3.

Even in the earliest stages of multiplication and division, the response to a problem like $24\div 6=?$ may well involve a procedure related to multiples of 6, in particular, $6\times 4=24$. In this example, we see that the question '24 divided by 6 is what?' or 'How many sixes are in twenty-four?' relates very closely to the fact that '6 multiplied by 4 is 24' or 'There are four sixes in twenty-four'. We can see that there is a number relationship that underlies both division and multiplication, in this case, it is the number triple (6, 4, 24).

In this chapter, I will consider how children experience the language of multiplication and division. First, I will review briefly some of the ways in which preliminary terms and concepts are introduced to the preschooler via everyday experiences. Then, the main part of the chapter will discuss ways of introducing the operations in school, pointing to some of the complexities and ambiguities inherent in these areas of early maths teaching, and identifying some common difficulties experienced by pupils. Some practical implications will be outlined.

Preliminary experiences in multiplication and division

Before children meet the arithmetic operations formally, it is crucial that they appreciate the nature of situations in which multiplication and division are inherent, that they understand the language of such situations and that they are familiar with procedures that will enable them to construct appropriate number triples. Before written problems involving the multiplication and division symbols are introduced, there are many practical ways in which children can gain insight into the meanings of these operations.

These include early linguistic experiences: much of the language associated with arithmetic operations is found in everyday activities in the home. For example, counting is developed in social interaction (Durkin *et al.*, 1986b) and through counting children learn to match number words and objects in one-to-one correspondence and, ultimately, to attach a single number word to a set of objects (Fuson, this volume). *Sharing* activities provide experience of the language and concepts of multiplication and division, exploiting terms such as 'lots of', 'each', 'equally' and 'fair' – crucial to later developments. Other important preliminary experiences arise in the course of activities such as climbing the stairs 'two at a time', introducing the notion of equal intervals on a number line and repetitive transactions (such as expenditure, where the child is introduced to issues of how many Xs may be purchased with a given amount of money). With these points in mind, let us consider how more efficient processes may be introduced at school.

Introducing the operations in school

The operations of multiplication and division are associated with many different actions. It is necessary for children to gain experience of a variety of situations that embody the operations before they are ready to generalize all of the many different processes as two single operations. Despite the dependence of both multiplication and division upon the same number triples, the operations are normally introduced independently. The following discussion reflects this established practice, but alternative procedures must also be considered.

Multiplication

A popular introduction for multiplication involves equal grouping and its associated language ('lots of' and 'sets of') as in Fig. 9.1. This is later formalized into '5 times 3' or '3 multiplied by 5' when the symbol for multiplication is introduced. The choice of language associated with the symbol is discussed later in this chapter.

An alternative visual image for multiplication is presented in the *array* where objects are arranged in *rows* and *columns*, as in Fig. 9.2a. There is an advantage in this representation, since the array illustrates '3 sets of 5' and at the same time '5 sets of 3'. This aspect is clear for adults but may call upon considerable experience for the children to appreciate the two distinct images within the same picture.

The language of arrays and the terms 'row' and 'column' may present

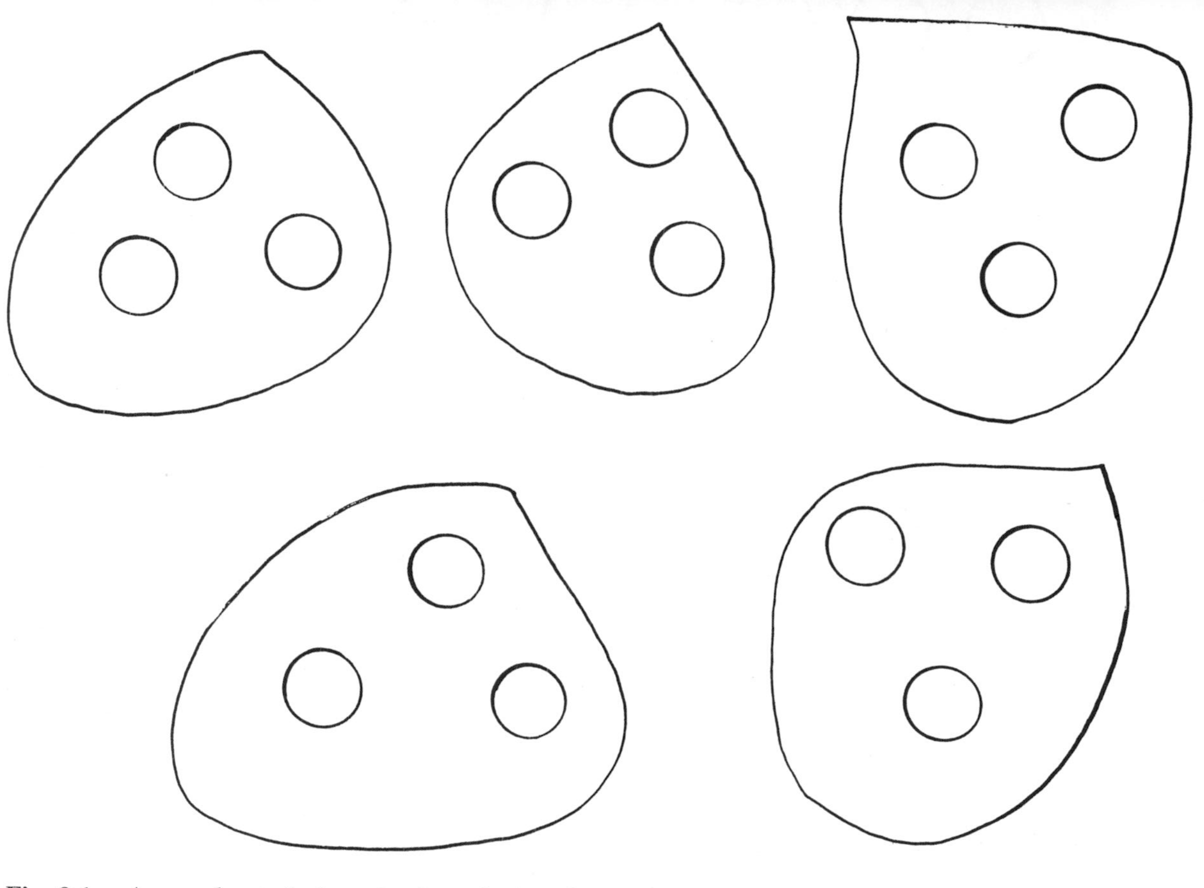

Fig. 9.1. A popular technique for introducing the concept of multiplication.

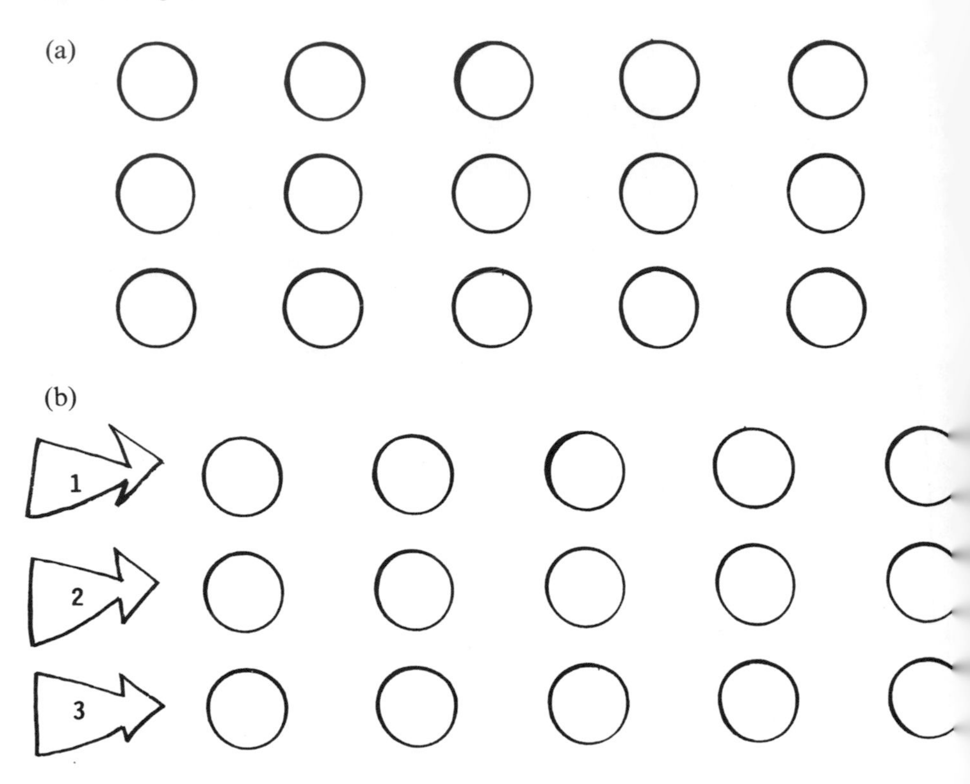

Fig. 9.2. Alternative techniques for illustrating multiplication.

difficulties, because the mathematical meaning differs from the everyday meaning. In the three-dimensional world, rows may be oriented in any horizontal direction, whereas columns are normally vertical. On two-dimensional paper, the 'horizontal' is restricted to the direction of lines of writing, the ortho (? dia)gonal direction being used to represent the vertical (see Durkin and Shire, this volume and Hanley, 1978, on the use of spatial terms in these contexts). A row of cars may extend *across* a car park or from the *front to the rear*, whereas columns are usually associated with *vertical supports* for buildings.

Confusion with the two-dimensional array may be exacerbated where a response to the question 'How many rows?' involves the indication of elements in the first column, as in Fig. 9.2b.

Division

Preliminary activities for division usually involve the actions of *sharing* (for which the mathematical term is *partition*) and *repeated subtraction* (for which the mathematical term is *quotition* or *measurement*). The idea of sharing some objects becomes synonymous with *dividing into equal portions*, through the actions involved with sharing a packet of sweets and cutting a cake relative to a set of discrete objects and a continuous whole, respectively. The formal operation of division is initially related only to creating equal subsets from a collection of discrete objects. It is very much later in the child's mathematical experience that the subdivisions of a whole are considered as fractions and later still that fractions are associated with the operation of division (see Kerslake, this volume). *Children may discuss their everyday experiences, yet be unable to formulate an appropriate mathematical expression.*

It is through concrete activities that children will meet variations in the language of division. Understanding precisely the implications of a variety of phrases will be essential if children are to interpret correctly word problems that they will meet later. The verb 'to share' relates to an activity of division, but the phrases 'shared by', 'shared among', 'shared between' and 'shared out' illustrate the subtle changes of phraseology that determine precisely the processes that are involved.

Remainders

In early experiences in the mathematics classroom, 'sharing' and 'repeated subtraction' situations are arranged so that the resulting 'portions' exhaust all the original collection. When children are familiar with this procedure, the idea of a *remainder* is introduced. The word 'remainder' is conventionally used to designate those elements that are 'left over' when portions of a given size have been constructed.

It is not a word that is commonly used by children (or anyone else!) and may prove quite demanding on the reading and writing skills of some children. The idea of creating a 'remainder' may not reflect the procedures that children will use in everyday experiences and will certainly not reflect the outcome of using a calculator for a division problem. It is only for a very limited period of the children's experience that the idea of a remainder will be appropriate.

Word problems in multiplication and division

When word problems are presented in multiplication and division (either through everyday experiences or in the mathematics classroom), the idea of equal groups or portions are often implicit and indicated by the word 'each'. Consider the example 'If Tom has 3 packets of marbles and each packet contains 15 marbles, how many marbles are there altogether?' The idea that this question involves 3 'lots of' or 'sets of' 15 marbles is indicated by the word 'each'. Research has shown that children do not always understand the implication of this word and may ignore it in their solution of problems.

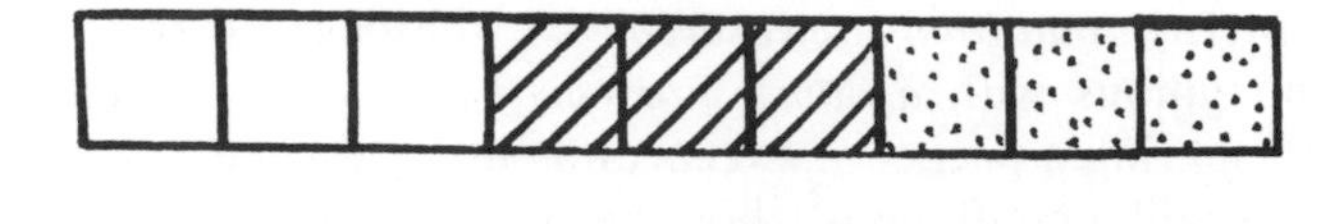

Fig. 9.3. Cubes as stimuli for tasks involving 'each'.

In an investigation of children's understanding of multiplication, Anghileri (1988) presented a task to children aged 5–8 years involving a stick of interlocking cubes in a pattern consisting of eight cubes, with two cubes each of four different colours. First, the children were asked preliminary questions to establish their understanding of the situation. They were asked how many different colours had been used to make the stick. Confusion arose when the children associated the question 'How many . . .' with a unitary count of each individual cube and responded with the total '8'. This question was followed by a question asking how many cubes of each colour had been used. The children's responses indicated that, again, some were confused and did not fully understand the question. When the children's responses to these two questions had been discussed with the interviewer so that, as far as possible, confusion had been sorted out, the children were asked to make a similarly constructed stick using '5 different colours with 3 cubes of each colour'. Examples like those shown in Fig. 9.3, consisting of three different colours with three cubes of each colour or five different coloured cubes together with three cubes of one colour, illustrate the children's difficulty in co-ordinating all the information given in the task and, in particular, some children appeared to have difficulty implementing the word 'each'.

Returning to the question specified above concerning Tom and his packets of marbles, additional confusion may arise with the use of the word 'altogether'. It has been shown (Nesher and Katriel, 1977) that children associate certain 'cue' words with the operations in arithmetic. In particular, the word 'altogether' is associated with addition and it is suggested that its use in the form '. . . how many altogether?' will lead children to assume that the operation involved must be addition.

In many word problems there are no 'cue words' to help children identify the appropriate operation. Take for example the problem 'My dog eats 4 biscuits every day. How long will a packet of 64 biscuits last?' Here there are no key words that will indicate that the operation is division. Children will meet many different types of word problems associated with multiplication and division. It is important that they are able to identify which operations are associated with each. Research has shown that high-school children have difficulty identifying multiplication and division as appropriate operations to solve word problems (Brown, 1981; Bell *et al.*, 1985; Fischbein *et al.*,1985). A more extensive discussion of word problems can be found in the chapters by De Corte and Verschaffel and by Teubal and Nesher (this volume).

Introducing the symbols

Any survey of school texts will soon reveal the complexity of language associated with the introduction of mathematical symbols for multiplication and division.

Multiplication

Where multiplication has been introduced as equal grouping, using either of the phrases 'sets of' or 'lots of', an intermediate notation for multiplication may be introduced through the use of brackets, thus:

4 sets of 3____4 (3)

4 lots of 3____4 (3)

When the multiplication symbol, 'x', is introduced, a decision must be made over the precise verbal interpretation that is to be used if ambiguity is to be avoided.

Within the adult and child population in the UK, many different interpretations of the multiplication symbol are known to exist (Anghileri, 1988):

- 3 times 4,
- 3 multiplied by 4,
- 3 multiply 4,
- 3 fours,
- 3 by 4,
- 4 threes,
- 3 lots of 4, and
- 3 timesed by 4.

Not all of these are considered to be acceptable and constructions like the last are discouraged. When consideration is given to each individual expression, ambiguities are revealed in their relationship to the symbolic expression '3×4'. The expression '3 times 4' refers to a set of 4 elements taken 3 times, while '3 multiplied by 4' refers to a set of 3 elements taken 4 times:

3 times 4 ________3 lots of 4 ______3(4) ________$4+4+4$

3 multiplied by 4 ______4 lots of 3 ______4(3) ________$3+3+3+3$

When the child understands the commutative rule for multiplication (that 3×4 and 4×3 will generate the same product), the discrepancy between the two expressions will be less important, but for the child struggling to attach some meaning to symbolic arithmetic expressions, the ambiguity may present a considerable barrier to understanding.

Division

A similar problem exists in division where a symbolic expression for 12 divided by 4 may be written in different forms, i.e. 12 ÷ 4, 4/12, $\frac{12}{4}$. The verbal expressions '12 divided by 4' and '4 divided into 12' are closely related. Brown (1981) found that many high-school children could not distinguish between symbolic expressions like '23 ÷ 291' and '291 ÷ 23', and when they were subsequently interviewed it was evident there was confusion between 'divided into' and 'divided by'. Children used both these expressions quite interchangeably, along with 'shared between'. The correct interpretation of the division '÷' symbol is 'divided by' and, although acceptable in some situations, the interpretation 'shared between' has only limited application to the division of whole numbers.

The idea of 'sharing' as an interpretation for division can cause difficulties in later school years when decimal numbers are involved (e.g. 0.4 ÷ 0.2), because the expression '0.4 shared between 0.2' has no meaning. Fischbein *et al.* (1985) suggest that high-school children may be inhibited in their selection of the correct operation to solve word problems involving division by the persistence of such interpretations.

Children's interpretation of symbolic expressions

Some interesting observations of children's understanding of the symbols for multiplication and division were made by McIntosh (1979) and Brown (1981). Brown asked children aged 11–13 years to 'make up a story' associated with an arithmetic expression involving addition, subtraction, multiplication and division (e.g. '9 × 3').

Multiplication was the most difficult operation for the children to illustrate, with only 53% of the 12-year-olds providing a multiplication story when the numbers were small, and 41% when large numbers were used. For division, a correct story was given by 69% for small numbers and by 56% for large numbers. The major difficulty appeared to be that of choosing the units for each number, e.g. where 'dogs' and 'people' were chosen, addition became difficult but multiplication was impossible. Examples given by the children include such stories as (McIntosh, 1979):

> Tim had 6 books × Mary had 3 books = 6 × 3.
>
> In the farmyard were 6 chickens and 3 pigs. My father said six times three is 18.

The idea of asking children to match a story to mathematical expressions involving the symbols for addition, subtraction, multiplication and division, has provided information on how they interpret the symbols and the situations they associate with each operation.

+ means 'and' or 'add'
– means 'take away'
÷ means 'share'
× means 'times'

The words 'add', 'take away' and 'share' are associated with concrete actions that may provide solution strategies for elementary problem solving but which inhibit understanding of applications in more advanced arithmetic. When decimal numbers are involved (e.g. $7.6-1.8$, $0.6\div 0.2$), Fischbein (1985) suggested that the 'primitive models' for arithmetic operations persist and conflict with the more complete mathematical notions necessary for understanding. The interpretation of the multiplication symbol as 'times' is not clearly associated with a concrete operation. This may account for the difficulty children experience in identifying multiplication as the appropriate operation in the solution of word problems (Brown, 1981; Bell, 1981; Fischbein, 1985). One may also consider the linguistic constructions associated with the phrases used.

Active and passive constructions

If one looks for an explanation of the popularity of the word 'times' in preference to 'multiplied by' and perhaps related confusion between 'divided into' and 'divided by', the discussion presented by Beilin (1975) suggests that *active* sentence forms (e.g. those involving 'times') are easier than *passive* sentence forms 'in both their comprehension and their production'.

The idea that children will use the 'active' construction 'Thomas hit me' well before they are able to construct the sentence 'I was hit by Thomas', is well known to most adults who have had the opportunity to observe the development of language in young children. Similarly, 3 times 4 may appear to children to be an easier construction to understand than 3 multiplied by 4. This, together with the prevalence of the use of the word 'times' by adults results in widespread use of the word by children. In an attempt to modify their language to fit the 'specifications' and 'expectations' of mathematicians, children will sometimes invent their own expressions like '3 timesed by 4', which is a compromise between the formal language of multiplication (3 multiplied by 4) and the expression they find easier to comprehend (3 times 4). Alternative constuctions used by adults and children remove the distinction between active and passive by modifying the expressions to '3 multiply 4', '3 by 4'. These alternative expressions for multiplication perhaps acknowledge the commutativity of multiplication whereby 3×4 and 4×3 are equivalent expressions which will result in the same product. Similar expressions for division do not exist as division is not a commutative operation.

Practical implications

The importance of multiplication and division means that we need to give careful thought to the ways in which children experience the language of these operations. Some key points arising from the present discussions are as follows.

1 Children can be prepared for the language of multiplication and division by means of a variety of early practical experiences. These include routinized activities in which number use is incidental but repetitive, as well as sharing games.
2 Teachers need to be prepared to allow children to use informal, everyday language to express their early ideas concerning multiplication and division that will be formalized mathematically at a later date.
3 We need to be aware that the child's perspective on a symbolic representation which appears to the adult to support the notions of multiplication and division may actually be quite different. For example, children may not see an array in terms of rows and columns as readily as the teacher does.
4 Evidence from word problems indicates that certain cue words can affect how children interpret the problem. We need to be aware of the possible consequences of the presence and absence of these words.

Conclusions

The language of multiplication and division presents many complexities to the young pupil, and their subtleties will take some time to acquire. In the process, the child will have to learn to express some tentative notions in new, increasingly formal terms. The child will have to cope with linguistic ambiguities and complex descriptions. Eventually, children will be introduced to formal mathematical terms such as *product*, *factor*, *division*, *multiplicand* and *dividend*. Guiding children from the earliest stages of mathematical operations through to the vernacular of the professional is a challenging task, but one which can be facilitated by careful attention to the language used.

10 Mixing metaphor and mathematics in the secondary classroom

Rita Nolder

Metaphor tends to be regarded as predominantly the concern of those involved in the study of literature, linguistics, philosophy and psychology, and mathematicians for the most part have ignored it. That is not to say that they have not made use of it, nor been influenced by it, but rather that they have not specifically acknowledged its role within mathematical experience. Many teachers of mathematics may be unaware that they are using metaphor in their classrooms and may regard it as irrelevant to their work since, in mathematics, everything is exactly what we say it is. However, metaphor permeates many aspects of mathematics teaching and learning.

In this chapter, I will consider some of the benefits and liabilities of uses of metaphor in mathematics education in the secondary years. First, I discuss briefly the definition of metaphor; secondly, I consider some spontaneous uses of metaphor by mathematics pupils and show how different metaphors can help or hinder progress on different tasks; thirdly, I turn to pedagogical uses of metaphor, which teachers sometimes exploit in attempts to facilitate their pupils' understanding. In the latter part of the chapter, I outline some practical implications, in terms of the possible value of metaphor in mathematics education, but indicating also some of the possible pitfalls. Overall, my theme will be that as metaphor is unlikely to disappear from the mathematics classroom, a better understanding of its functions and an increased awareness of its advantages and disadvantages could lead to more effective ways of teaching and learning mathematics.

Defining metaphor

So what is metaphor? Metaphor has been defined by Low (1988: 126) as a reclassification which involves 'treating *X* as if it were, in some ways, *Y*', where *X* and *Y* have been described as the primary and secondary subjects (Black, 1979:

28), respectively, of the metaphor. The choice of *Y* is based on a point, or points, of resemblance between *X* and *Y*, although the expression formed is usually meaningless if taken literally. A metaphor may be expressed in the form '*X* is *Y*' as in such mathematical metaphors as:

- A function is a machine
- An equation is a balance
- Algebra is shorthand
- Prime numbers are primary colours

In other metaphorical expressions, e.g. the 'skeletons' of polyhedra or the 'rules' of algebra, the secondary subject may initially be less obvious. Within this chapter, I shall adopt an interaction view of metaphor which 'approaches metaphor functionally rather than grammatically' (Ortony, 1979: 923–4). Thus, although the grammatical structure of a statement may be that of a simile, '*X* is like *Y*', rather than of a standard metaphorical form, such as '*X* is *Y*', it is the thought process which underlies the statement which determines whether or not it is classified as a metaphor.

Metaphor is a basic feature of human communication. It enables people to deal with novel experiences, whether describing something which they have never previously encountered, or seeking to comprehend a new idea. It has been suggested that 'we examine the unknown, scanning it over and over until we can describe it in terms of the known' (Sutton, 1978: 11). Phrases such as 'It's like . . .' or 'It's as if . . .' spring to mind.

Pupils' uses of metaphor in mathematical tasks

The following examples from several mathematics classrooms (Nolder, 1984) illustrate the human predilection for asserting likeness, which is at the heart of the metaphoric process.

> A seventeen year-old describes one graph as a "U-shaped valley" whereas another is a "hump back bridge".
>
> A low-achieving fourteen year-old describes a parallelogram as "like a pushed-over rectangle".
>
> $(A')' = ?$ Well, it's like $-(-3)$ isn't it? That's $+3$. It's like double negatives in English too. So I think $(A')'$ must be A.
>
> Michelle suddenly noticed that complex numbers were "like surds", checked on how to divide by surds which she only dimly recalled, applied the same method to complex numbers and found that it worked.

The first two examples illustrate a use of metaphor which is essentially comparative, aiding communication by enabling the speaker to describe a new mathematical idea in terms of something more familiar, whether from within or outside the field of mathematics. It is this particular use of metaphor which has helped mathematical language to expand, as words are introduced to fill gaps in

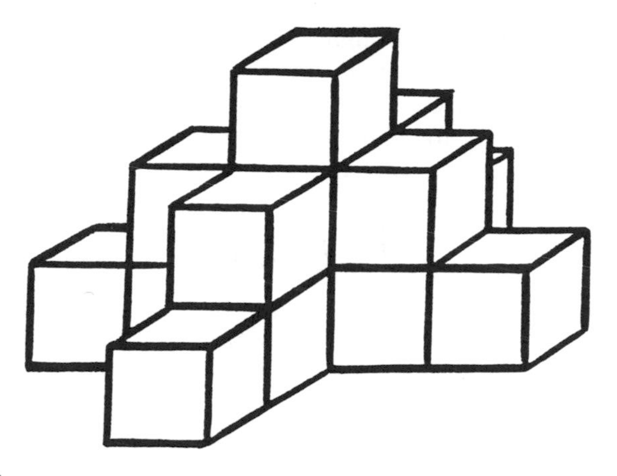

Fig. 10.1. The tower.

the mathematical lexicon because of some perceived similarity between the mathematical meaning and everyday meaning. Examples of this are 'tree' diagrams, 'similar' shapes, 'rational' and 'irrational' numbers, the 'domain' and 'range' of a function, the 'neighbourhood' of a point, the 'nodes' of a network, etc. When a speaker has been familiar with such expressions for many years, their metaphoric quality may go unnoticed. However, for a teacher it is important to explore pupils' understanding of such metaphoric terms to find out what meanings, if any, these evoke, and to clarify any ambiguity arising between the mathematical meaning and the ordinary English meaning.

The remaining two examples of metaphors produced by pupils entail more than a simple comparison. In each case, the comparison results in the emergence of new insights relating to the primary subject of the metaphor. Hence the perceived similarity between 'complementing a complement' in set theory and 'double negatives', whether in mathematics or English, leads to what is known about double negatives suggesting a rule for complements of complements. In the same way, the perception of a similarity between complex numbers of the form $a+b_i$ and surds of the form $a+\sqrt{b}$ leads to the transfer of the method for dividing surds to an analogous one for dividing complex numbers.

In essentially the same situation, individuals may generate different metaphors when describing the same mathematical object or idea and, as the following example illustrates, some metaphors may be more helpful to the learner than others.

A mathematical problem which is often posed to secondary school pupils involves predicting the number of cubes required to produce a tower similar to one which they have been given (Fig. 10.1) but which is n rather than three cubes high.

The most common metaphors spontaneously produced to describe the four identical configurations of cubes placed symmetrically around the centre column are 'staircases', 'wings' and 'triangles'. All these terms help the learner to communicate a method for calculating the number of cubes in the original tower. Such expressions are usually of the form 'You multiply the number of cubes in a staircase/wing/triangle by four and add on the number of cubes in the centre column.' However, when encouraged to find a more efficient way to calculate the number of cubes, since counting cubes for a tower which is perhaps 1000 cubes high is time-consuming, the 'staircase' and 'wing' metaphors prove less helpful than the 'triangle' metaphor. The latter metaphor, with its associated fact that a triangle is half a rectangle, may lead to pupils forming two rectangles measuring $(n-1)$ by n, plus a central column of height n (Fig. 10.2). From here the algebraic generalization $2 \times n \times (n-1)+n$ for the number of cubes needed for a tower of height n *is* produced.

Other pupils confronted with this problem have 'seen' the tower as something else, either from the initial stages or part way through the problem. The 'something else' usually takes the form either of one rectangle (Fig. 10.3) or a square and a rectangle (Fig. 10.4).

Teachers' use of metaphor in mathematics

The focus so far has been on the metaphors produced spontaneously by pupils in the course of their mathematical activity. Next we turn to what may be described as pedagogical metaphors, those used explicitly by teachers to facilitate their pupils' understanding of mathematical concepts. The way in which such metaphors work may be illustrated by considering a common and well-established metaphor, now enshrined in numerous secondary school texts, an equation is a balance.

For what reason does a teacher choose to use this metaphor, since it is possible to teach someone to solve linear equations without recourse to imagery? Consciously, or unconsciously, the teacher is offering pupils something concrete and familiar to help them to understand an unfamiliar, abstract idea. At the same time, the teacher is linking a new concept to the learner's past experience. By making links between a new concept and a well-established, if not mathematical one, the teacher aims to make that piece of mathematics more secure and hopes that the metaphor by its novelty may enhance memorability.

If we analyse the metaphor an equation is a balance, then the primary subject is the equation and the balance is the secondary subject. The two subjects 'interact' as follows. The presence of the primary subject causes the hearer of the metaphorical statement to stress some features of the secondary subject and to ignore others. The stressed features form the 'secondary implicative complex' (Black, 1979: 29). For the metaphor an equation is a balance, members of the secondary implicative complex to which teachers might direct their pupils' attention include:

(A) A balance has two scale pans and a beam.
(B) You put weights on to the pans.

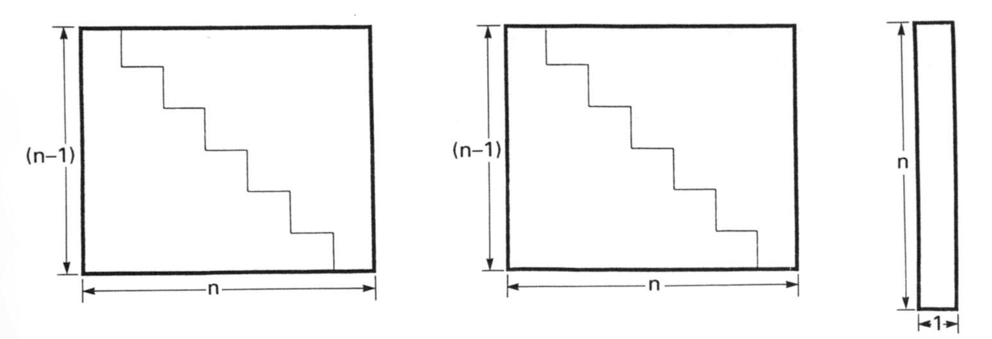

Fig. 10.2. Pupil's reconfiguration of the tower (1).

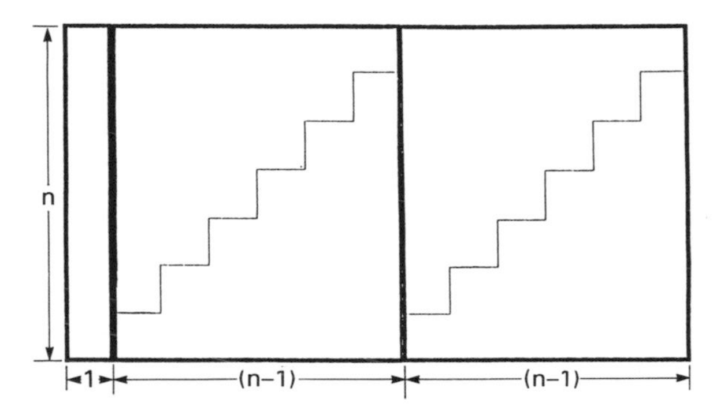

Fig. 10.3. Pupil's reconfiguration of the tower (2).

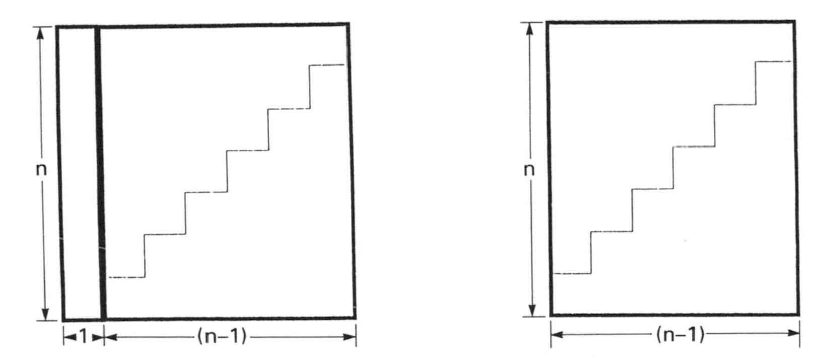

Fig. 10.4. Pupil's reconfiguration of the tower (3).

(C) For equilibrium, the weight on one side must be the same as the weight on the other.
(D) If you add/remove weights to/from one side you must add/remove weights to/from the other, to preserve balance.

Pupils are then encouraged to construct a parallel implicative complex which fits the primary subject. Corresponding members of the primary implicative complex would be

(a) An equation has two sides separated by an '=' sign.
(b) You put numbers and/or letters on both sides.
(c) For equality, the value of the numbers and/or letters on both sides must be the same.
(d) What you add to/subtract from one side you must add to/subtract from the other to preserve equality.

In constructing the analogy between the two implicative complexes, changes may be induced in the secondary implicative complex. In the metaphor an equation is a balance, because weights have been identified with numbers, what we know about numbers (e.g. that multiplication is repeated addition,) enables us to modify a member of the secondary implicative complex. D and d then become:

(D′) What you do to the weights on one side you must do to the weights on the other.
(d′) What you do to one side of the equation you must do to the other.

The effectiveness of such pedagogical metaphors in aiding pupils' understanding is based on their familiarity with the secondary subject, and without this the metaphor may well be a pedagogical disaster.

Even when pupils do have a good acquaintance with the secondary subject, in this case balances, they may still experience some difficulty when expected to apply what they know about balances to equations because of the existence of aspects of the secondary subject which are not analogous to the primary subject. For example, a pupil confronted with a textbook illustration involving cans resting on the scale pans of a traditional beam balance, was perturbed by the positioning of the cans. She knew from practical experience that this would not work on a real balance because of the need to position weights very carefully to preserve equilibrium. This discrepancy distracted the pupil from the concept being developed and it required skilful handling by the teacher to remove the confusion.

Metaphor and misunderstanding in mathematics

The example of the balance metaphor draws attention to the fact that while metaphors may be of pedagogical value, they can also engender misunderstanding and confusion in the mathematics classroom. A second example, algebra is shorthand, points to further problems inherent in teachers' use of metaphor.

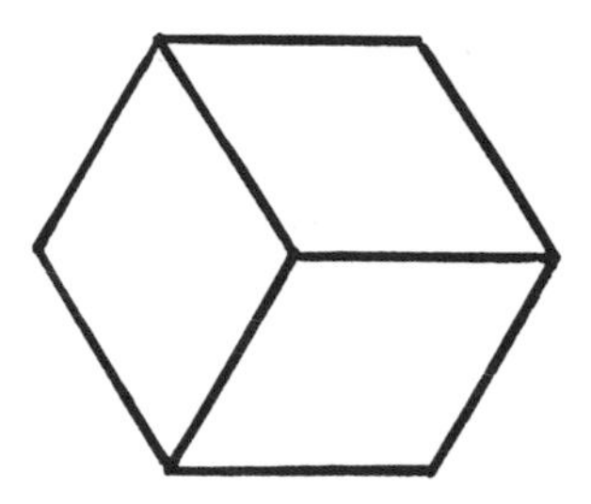

Fig. 10.5. Rhombuses, or a cube, or a square?

It is common for teachers to suggest that letters are short forms of words, and the identification a=apples, b=bananas, etc., is often made so that an expression such as $2a+3b$ means two apples and three bananas. While such identification may prove helpful as a starting point, there are dangers inherent in its use. It obscures the mathematical meaning of $2a+3b$ which is 'two times the number of apples plus three times the number of bananas', where a and b represent the numbers of apples and bananas, respectively. In the metaphoric identification of letters and objects, the concept of variable is not only missed but its development may be hindered. The results of one research programme (Hart, 1981) indicate that many pupils do not progress beyond the stage of 'letter as object' in their understanding of algebra.

Misunderstandings do not only arise from teachers' metaphors. We saw in earlier examples that pupils' ability to see one thing as if it were something else, can prove helpful in the course of mathematical activity; however, it may on occasions give rise to confusion. Such confusion often manifests itself within pupils' mathematical errors, and its source may not surface unless pupils are encouraged to talk through what they are thinking. An example of confusion arising from 'seeing as' arose in a class of 12-year-olds (Nolder, 1984: 79–80). The pupils were naming the polygons which they could see in posters of stained glass. One poster was composed of hexagons and rhombuses, the predominant design being of the form in Fig. 10.5.

One pupil stated confidently that he could see a square. The first reaction of the other pupils was laughter, then one after another they too began to 'see' the square. After some laboured explanation by the class, the teacher came to see the square as well. What they saw was the configuration of rhombuses 'as a cube' and, because a cube is 'like a square' and they were dealing with two-dimensional shapes, a square was what was required. Jaworski (1988: 287–8) has reported a similar experience, also with a class of 12-year-olds.

Practical implications

Developmentalists have shown the acquisition of metaphorical competence to be a protracted affair, extending through middle childhood (cf. Gardner *et al.*,

1978), but that it is reasonable to expect secondary age children to be able increasingly to handle concepts presented in this way. However, as the examples in this chapter illustrate, teachers need to be sensitive to the occurrence of metaphors and to the potential advantages and disadvantages to pupils' learning involved in their use in the mathematics classroom.

Why metaphor is useful

Low (1988) has outlined various benefits of the uses of metaphor in language teaching, and several of these apply equally to aspects of mathematics education that I have considered in this chapter:

1 *Metaphor makes it possible to talk about X at all.* As we have seen, metaphorical descriptions enable us to address phenomena that are otherwise difficult to describe, such as the tower in Fig. 10.1, or the nature of a function.
2 *Metaphor makes it possible to relate new concepts systematically to things already understood.* The metaphor a complex number is a vector, for example, helps students who already have a knowledge of vectors to grasp aspects of complex numbers, since comparisons between x- and y-components and real and imaginary parts, magnitude and modulus, direction and argument are easily made.
3 *Metaphor extends thought.* The pupil who perceived an analogy between complex numbers and surds was thus enabled to extend her concept of complex numbers.
4 *Metaphor compels attention.* Because metaphors seldom make sense when taken at face value, they challenge the hearer to search for the inherent meaning. The metaphor prime numbers are primary colours is a striking example of this aspect of metaphor.

Teachers and pupils can and do employ metaphor in these and related ways to facilitate mathematical understanding.

Problems incurred by the use of metaphor

We need also to be aware of liabilities in the use of metaphor, some of which have already been mentioned in earlier sections of this chapter. There are several points to be borne in mind here.

1 *Metaphors can blinker or distort perception.* For example, the confusion engendered by a pupil seeing a two-dimensional design featuring rhombuses as a cube and then reinterpreting this as a square has already been mentioned.
2 *Some elementary metaphors fail to apply in more advanced contexts.* For example, as Kerslake (this volume) elaborates, children who were once taught the rule that 'multiplication makes bigger' are baffled when they come to use it

in the context of fractions. The rule does work when the natural numbers are identified with fractions of unit denominator, but cannot be extended to the rest of the set of fractions.

3 *One metaphor may not adequately structure a concept.* This is true of many metaphors (cf. Lakoff and Johnson, 1980), not only in a mathematical context. Within mathematics we have noted, for example, that common difficulties in algebra may result from the way in which pupils are encouraged to see algebra as shorthand. This does not provide an adequate structuring of the concept and may retard the development of the idea of letters as variables.

4 *Metaphors may be over-extended beyond their utility.* Referring again to a complex number is a vector, students may extend the analogy still further by referring to the 'resultant' when adding two or more complex numbers (Nolder, 1985: 34).

Conclusions

Consideration of the various mathematical metaphors we have encountered in this chapter draws attention to the need to clarify the meanings which we and our pupils attach to the numerous metaphorical utterances in evidence in the mathematics classroom. In addition, an exploration of the limitations of these metaphors is essential if we are not to produce confused and discouraged pupils.

We have looked at a limited number of examples of pupils' metaphors, teachers' metaphors and metaphors inherent in mathematics itself. In so doing, we have only touched the tip of the iceberg. It has not been the purpose here to advocate or decry the use of metaphor in the mathematics classroom, but rather to draw attention to its existence. If teachers and pupils alike were to become more aware of the 'Why?', 'What?' and 'How?' of metaphor-making, then perhaps metaphor might make a more effective contribution to the secondary mathematical experience.

Further reading

Pimm, D. (1987). *Speaking mathematically*, ch. 4. London Routledge.
Vosniadou, S. (1987). Children and metaphors. *Child Development*, **58**, 870–85.

Section 4

Word problems

11 Some factors influencing the solution of addition and subtraction word problems

Erik De Corte and Lieven Verschaffel*

Introduction

'Ann and Tom have 8 books altogether; Ann has 5 books; how many books does Tom have?' For a long time, word problems such as the above have constituted an important part of mathematics education in the primary school. Their role dates back to antiquity. Indeed, one can find verbal problems in 4000-year-old Egyptian papyri. They also figure in Greek and Roman manuscripts as well as in arithmetic textbooks from the early days of printing. On the other hand, the international literature on mathematics education shows that, despite this long tradition, a large number of children and teachers find word problems very difficult to learn and to teach, respectively. Our contacts over the past few years with superintendents, principals and teachers of Flemish primary schools, confirm the 'word problem depression' in educational practice.

However, it seems to us that there is some hope for improvement. Indeed, a large amount of research undertaken since the late 1970s in different parts of the world has resulted in a substantial series of findings and principles that provide a relatively firm basis for revising different aspects of learning and instruction of word problems. In this chapter, we present some of these findings, focusing on the effects of three distinct task characteristics on children's solutions of elementary addition and subtraction word problems; namely (1) the nature of the semantic structure underlying the problem, (2) the extent to which this semantic structure is made clear and explicit in the problem text, and (3) the order of presentation of the known sets of the underlying semantic structure. Although we will also refer regularly to the work of other researchers, our discussion of the influence of these task characteristics on children's solutions will be based on our own investigations since the early 1980s. Therefore, we will first provide some background information concerning our approach and methodology.

*L. Verschaffel is a research associate of the National Fund for Scientific Research, Belgium.

Our research on young children's problem solving is strongly influenced by the information-processing approach of cognitive psychology (e.g. Newell and Simon, 1972; Tuma and Reif, 1980) and by the work of Greeno and associates in particular (Riley *et al.*, 1983). Based upon the early studies of those scholars on the one hand, and on the results of our own empirical investigations on the other, we have developed a competent problem-solving model of elementary arithmetic word problems comprising five stages (De Corte and Verschaffel, 1985a; Verschaffel, 1984).

1 A complex, goal-oriented, text-processing activity occurs: starting from the verbal text, the pupil constructs a global abstract, internal representation of the problem in terms of sets and set relations.
2 On the basis of that representation, the problem solver selects an appropriate formal arithmetic operation or an informal counting strategy to find the unknown element in the problem representation.
3 The selected action or operation is executed.
4 The problem solver reactivates the initial problem representation, replaces the unknown element by the result of the action performed, and formulates the answer.
5 Verification actions are performed to check the correctness of the solution found in the preceding stage.

Diverse research methods have been used, which include collective paper-and-pencil tests administered to large groups of children (100–200 pupils), videotaped individual interviews with smaller groups of children (20–30 pupils), studies of eye-movements during problem solving, textbook analysis and teaching experiments. Using this variety of research techniques in a series of studies guided by the theoretical model of competent problem-solving described above, we obtained the findings reported in this chapter.

Research findings

The effects of semantic structure on children's solutions of elementary arithmetic word problems

Most older studies of the effects of task variables on children's arithmetic word problems solving have concentrated on the surface characteristics and on the mathematical structure of those problems. Some typical variables examined include the number of words in the problem, the structure of the number sentence 'hidden' in the problem, the place of the question, the presence of a cue or key word, etc. (Verschaffel, 1984). In more recent work, the focus has shifted from mathematical and surface aspects towards the semantic structure of the problem. This new approach is based on two assumptions:

1 Word problems that are solvable using the same arithmetic operation, can be

described in terms of different networks of concepts and relationships underlying the problem.
2 Constructing an appropriate internal representation of such a conceptual network is a crucial aspect of expertise in word problem solving (see stage 1 of our competent word problem solving model).

In the late 1970s, this approach was first applied to simple addition and subtraction word problems. Greeno and associates (Riley *et al.*, 1983) introduced a classification scheme in which they distinguished three semantic categories: 'change', 'combine' and 'compare' problems:

- *Change problems* refer to dynamic situations in which some event changes the value of a quantity, e.g. 'Joe had 3 marbles; Tom gave him 5 more marbles; how many marbles does Joe have now?'
- *Combine problems* relate to static situations involving two amounts, that are considered either separately or in combination, as in the following example: 'Joe has 3 marbles; Tom has 5 marbles; how many marbles do they have altogether?'
- *Compare problems* involve two amounts that are compared and the difference between them, e.g. 'Joe has 3 marbles; Tom has 5 more marbles than Joe; how many marbles does Tom have?'

Each of these three categories can further be subdivided in distinct problem types depending on the identity of the unknown quantity; for change and compare problems, further distinctions can be made depending on the direction of the event (increase or decrease) or the relationship (more or less), respectively. Combining these three characteristics, Greeno and associates distinguished 14 types of simple addition and subtraction word problems (see Table 11.1).

Robust research evidence is now available which shows the psychological significance of the semantic classification of word problems. A major finding in this respect derives from children's performance on such tasks: *word problems that can be solved by the same arithmetic operation but differ with respect to their underlying semantic structure have very different degrees of difficulty*.

As an illustration, Table 11.2 shows the results from a study in which a large group of second-graders was given a series of 14 simple addition and subtraction word problems representing the distinct types from the classification scheme presented in Table 11.1 (Pauwels, 1987). The combine 1 problem, the change 6 problem and the compare 3 problem can all be solved by the same arithmetic operation, namely adding the two given numbers. However, Table 11.2 shows that their levels of difficulty differed strongly: 97, 83 and 47% correct, respectively. The same holds for subtraction problems: the change 2, the combine 2 and the compare 5 problems can all be solved by decreasing the larger given number by the smaller one; nevertheless, they were solved correctly by 88, 62 and 46% of the children, respectively.

According to our theoretical model, the difficulty level of semantically distinct problems can differ either because the semantic schemes necessary to represent

Table 11.1 Different types of elementary addition and subtraction word problems (Riley *et al.*, 1983)

Type	*Example*	*Schema*	*Direction*	*Unknown*
Change 1	Joe had 3 marbles; then Tom gave him 5 more marbles; how many marbles does Joe have now?	change	increase	result set
Change 2	Joe had 8 marbles; then he gave 5 marbles to Tom; how many marbles does Joe have now?	change	decrease	result set
Change 3	Joe had 3 marbles; then Tom gave him some more marbles; now Joe has 8 marbles; how many marbles did Tom give him?	change	increase	change set
Change 4	Joe had 8 marbles; then he gave some marbles to Tom; now Joe has 3 marbles; how many marbles did he give to Tom?	change	decrease	change set
Change 5	Joe had some marbles; then Tom gave him 5 more marbles; now Joe has 8 marbles; how many marbles did Joe have in the beginning?	change	increase	start set
Change 6	Joe had some marbles; then he gave 5 marbles to Tom; now Joe has 3 marbles; how many marbles did Joe have in the beginning?	change	decrease	start set
Combine 1	Joe has 3 marbles; Tom has 5 marbles; how many marbles do they have altogether?	combine	—	superset
Combine 2	Joe and Tom have 8 marbles altogether; Joe has 3 marbles; how many marbles does Tom have?	combine	—	subset
Compare 1	Joe has 8 marbles; Tom has 5 marbles; how many marbles does Joe have more than Tom?	compare	more	difference set
Compare 2	Joe has 8 marbles; Tom has 5 marbles; how many marbles does Tom have less than Joe?	compare	less	difference set

Table 11.1 (*Continued*)

Type	*Example*	*Schema*	*Direction*	*Unknown*
Compare 3	Joe has 3 marbles; Tom has 5 more marbles than Joe; how many marbles does Tom have?	compare	more	compared set
Compare 4	Joe has 8 marbles; Tom has 5 marbles less than Joe; how many marbles does Tom have?	compare	less	compared set
Compare 5	Joe has 8 marbles; he has 5 more marbles than Tom; how many marbles does Tom have?	compare	more	reference set
Compare 6	Joe has 3 marbles; he has 5 marbles less than Tom; how many marbles does Tom have?	compare	less	reference set

Table 11.2 Percentages of correct solutions for the 14 types of word problems in Pauwels' (1987) study ($N = 126$)

Problem type	*% correct*
Change 1	97
Change 2	88
Change 3	73
Change 4	78
Change 5	67
Change 6	83
Combine 1	98
Combine 2	62
Compare 1	79
Compare 2	78
Compare 3	47
Compare 4	44
Compare 5	46
Compare 6	38

Table 11.3 Material solution strategies on subtraction problems (Verschaffel, 1984)

Problem type	*Example*	*Solution strategy*
Change 2	Pete had 6 apples; he gave 2 apples to Ann; how many apples does Pete have now?	*Separating from*: using objects or fingers, the child constructs a set corresponding to the larger given number (6) in the problem; then removes as many objects as indicated by the smaller number (2); the answer is the remaining number of objects (4)
Change 3	Pete had 3 apples; Ann gave him some more apples; now Pete has 10 apples; how many apples did Ann give to Pete?	*Adding on*: the child constructs a set corresponding to the smaller given number (3); then adds elements to this set until there are as many objects as indicated by the larger number (10); the answer is found by counting the number of objects added (7)
Compare 1	Pete has 3 apples; Ann has 8 apples; how many apples does Ann have more than Pete?	*Matching*: the child constructs a set corresponding to the smaller (3) and a set corresponding to the larger given number (8) and matches them until one set is exhausted; the answer is the number of objects remaining in the unmatched set (5)

the different problem types are not equally well mastered by children, or because some problem representations are more easily mapped with an appropriate arithmetic operation than others.

A second main finding supporting the importance of the semantic distinctions concerns the relationship between the strategies children use to solve word problems and their underlying structure. Indeed, our data indicate that problems solvable with the same arithmetic operation, but differing in their semantic structure, elicit different solution strategies. To illustrate this, we mention a finding concerning children's material solution strategies for subtraction problems, obtained in a study in which 30 first-graders were individually given a series of simple addition and subtraction work problems three times during the school year (Verschaffel, 1984). We observed a significant trend to apply different material solution strategies for distinct types of subtraction problems. More specifically, they tended to solve each subtraction problem with that kind of strategy that corresponds most closely to its semantic structure. Table 11.3 gives an overview of the most frequently occurring material strategies on three subtraction problems in our study.

As we have reported in detail elsewhere (De Corte and Verschaffel, 1987), the relationship between the semantic structure of elementary arithmetic word problems on the one hand, and children's solution strategies on the other, does not only hold for pupils solving problems with the help of concrete objects such as fingers or blocks, but also for those applying verbal counting strategies (based on forward or backward counting) or even mental solution strategies (based on recalled number facts). For example, in the above-mentioned study, it was also found that combine 1 problems ('Pete has 3 apples; Ann has 7 apples; how many apples do they have altogether?') elicited much more verbal and mental solution strategies beginning with the second given number (e.g. '7 . . . , 8, 9, 10') than strategies starting with the first one (e.g. '3 . . . 4, 5, 6, 7, 8, 9, 10'). However, for the change 1 problem ('Pete had 3 apples; Ann gave Pete 5 more apples; how many apples does Pete have now?'), we found the opposite tendency: most children started their solution strategy with the smaller given number, namely 3. In terms of our theoretical model, this finding can be explained as follows. Mapping a mental representation of an addition problem starting with the smaller of the two given numbers with a solution strategy beginning with the larger one requires the child to interchange the two given quantities in his or her problem representation. This reorganization of the original problem representation is easier to do when the semantic structure does not involve temporal sequence (as in combine problems), than when there is such sequence of events described in the problem (as in change problems).

Clarity of the problem statement

Although the semantic structure appears to be a major factor determining problem solution, recent research has convincingly shown that other task characteristics can also significantly affect children's performance and strategies with respect to verbal problems. In this chapter, we will review two additional task characteristics, namely the degree to which the underlying semantic structure is made explicit in the problem text and the order of presentation of the given numbers. In this section, the first aspect is discussed.

Consider, for example, the following example: 'Ann and Tom have 8 books altogether; Ann has 5 books; how many books does Tom have?' In this problem text, it is not explicitly stated that Ann's 5 books mentioned in the second sentence, are at the same time part of the 8 books that Ann and Tom have altogether. But this combine 2 problem can be reworded in such a way that its surface structure makes the semantic relations more obvious: 'Ann and Tom have 8 books altogether; 5 of these books belong to Ann and the rest belong to Tom; how many books does Tom have?' In his individual interviews, Verschaffel (1984) observed that some children who could not solve the standard combine 2 problem, answered the reformulated version correctly. Using similar words, Carpenter (1985) showed that subtle aspects of the formulation of the problem, such as the tenses of the verbs in the problem text, may be responsible for observed differences in difficulty between variants of change problems.

Furthermore, Hudson (1983) found that young children presented with a picture (e.g. showing five birds and four worms), performed differently depending on the statement of the question. More specifically, performances were much worse on the question 'How many more birds than worms are there' than on an alternative question 'Suppose the birds all race over and each one tries to get a worm, how many birds won't get a worm?'; in the latter case, most children used a matching strategy to obtain their solution. Hudson's (1983) data suggest that for compare problems too, children's difficulties are influenced by the formulation of the problem.

Based on these observational and experimental data, we carried out an investigation in which we tested systematically the hypothesis that rewording simple addition and subtraction problems in such a way that the semantic relations are made more explicit without affecting their semantic structure, facilitates the solution of these problems by young elementary school children (De Corte *et al.*, 1985a). Two series of six rather difficult word problems – two combine 2, two change 5 and two compare 1 problems – were administered collectively near the end of the school year to 89 first- and 84 second-graders. In series A, the problems were stated in the usual condensed form; in series B, they were reformulated as in the above-mentioned studies. The results supported our hypothesis: in each of the three semantic structures, the reworded problems of series B were solved significantly better than the standard problems of series A.

Our explanation of these findings is as follows. The mental representation constructed in the first stage of a *competent* solution process, is considered the result of a complex interaction of top-down and bottom-up analysis; that is, the processing of the verbal input as well as the activity of the competent solver's semantic schemes contribute to the construction of the representation. In less able and inexperienced children, the semantic schemes are not yet very well developed; so these children depend more on bottom-up or text-driven processing to construct an appropriate problem representation. Therefore, we hypothesize that – especially for less able children – rewording verbal problems so that the semantic relations are made more explicit without affecting the underlying semantic structure, facilitates the construction of a proper problem representation.

Order of presentation of the given numbers

Another task characteristic that significantly influences children's solution processes, is the sequence of the given numbers in the problem text. Verschaffel (1984) found that children solved combine 2 problems almost exclusively with a so-called *indirect additive* (IA) strategy, either adding on (when using blocks) or counting up from the smaller given number (when working at the verbal counting level). Carpenter and Moser (1984), on the other hand, reported that the majority of the children in their study applied a *direct subtractive* (DS) strategy to solve combine 2 problems, either separating from or counting down from the larger given number.

However, a comparison of the combine 2 problems used in both studies revealed a significant difference. In Verschaffel's (1984) problem, the larger number was mentioned first ('Pete has 3 apples; Ann has also some apples; Pete and Ann have 9 apples althogether; how many apples does Ann have?'); in Carpenter and Moser's (1984) problem, the sequence of the two given numbers was reversed ('There are 6 children on the playground; 4 are boys and the rest are girls; how many girls are there on the playground?). This suggests the following hypothesis, which we have tested in a separate study: the strategies children use to solve simple addition and subtraction problems depend not only on the semantic structure underlying the task (see p. 118), but also on the sequence of the given numbers in the problem text.

A test was constructed containing 20 elementary addition and subtraction word problems representing four different types from the classification schema of Riley *et al.* (1983): change 1, combine 1, change 3 and combine 2. For each problem type, several variants were constructed in which the order of presentation of the given numbers was systematically varied (Verschaffel and De Corte, 1990). This test was administered collectively in the beginning of the school year to 85 second-graders. They were not only asked to solve the problems, but also to write down for each problem a number sentence showing which arithmetic operation had been performed to find the solution. Near the end of the next school year, a smaller group of first-graders was individually interviewed using exactly the same set of word problems. Each problem was read aloud by the interviewer and the children were asked to solve it and to explain their solution strategy. We will discuss the results for one addition (change 1) and one subtraction (combine 2) problem.

With regard to addition problems, distinction can be made between strategies in which the child begins with the first of the two given numbers (F strategies) and strategies starting with the second one (S strategies) (examples of both kinds of strategies can be found on pp. 122–23). For problems in which the first given number is the smaller one, S strategies are more efficient; indeed, by disregarding the given order of the addends and starting with the larger one, the number or difficulty of the steps in the solution is reduced to a minimum. As expected, S strategies were applied much more frequently for change 1 problems beginning with the smaller given number (48%) than when the larger number was given first (14%).

With respect to subtraction problems, we were especially interested in the influence of the order of presentation of the two given numbers on the choice of either a DS or an IA strategy. In DS strategies, the answer is found by subtracting the smaller given number from the larger one; in IA strategies, the child determines what quantity the smaller given number must be added to, to obtain the larger one. The results show that combine 2 problems starting with the smaller and the larger given number elicited considerably different percentages of DS and IA strategies: whereas for the former most children (82%) applied an IA strategy, the latter elicited a relatively high proportion of DS strategies (43%). This implies that the order of presentation of the two given sets indeed plays an important role in children's solution strategies for combine 2 problems.

The results of this study are not in conflict with the well-documented finding concerning the effects of the semantic structure on children's solutions of elementary arithmetic word problems, but rather complementary. Indeed, our data show that with respect to young problem solvers, considerable differences in solution strategies can occur within a given semantic problem type, depending on the order of presentation of the sets in the problem text. This finding, too, should be taken into consideration when constructing elementary arithmetic word problems. Further discussion of the relationship between the order of items in a word problem and the order of the arithmetical steps involved, is presented in the following chapter by Teubal and Nesher.

Practical implications: strategies for instruction

In the preceding section, we reported a series of recent research findings concerning the effects of a number of task characteristics on children's solutions of elementary addition and subtraction word problems, namely (1) the nature of the semantic structure underlying the problem, (2) the extent to which this semantic structure is made clear and explicit in the problem text, and (3) the order of presentation of the given numbers.

It is interesting to relate these findings to current instructional practice, as it emerged from our analysis of a representative sample of six Flemish textbooks for mathematics education in the first grade (De Corte *et al.*, 1985b).

First, we found a remarkable one-sidedness in the word problems used in current educational practice when compared to the 14 categories listed in Table

11.1. In most textbooks, there was a substantial preponderance of change 1 and 2 (e.g. 'Pete had 6 apples; Ann gave Pete 2 more apples; how many apples does Pete have now?') and combine 1 problems (e.g. 'Pete has 3 apples; Ann has 5 apples; how many apples do they have altogether?'). In several textbooks, very few or no problems of the following types occurred: change 3, 4, 5 and 6 and combine 2. In three out of the six textbooks, not a single compare problem was found; in two textbooks there were a few compare problems, while one had a relatively good balance between the three semantic types. Stigler *et al.* (1986) have reported a similar one-sidedness in an analysis of four widely used American elementary mathematics textbooks. Taking these findings into account, it can be recommended that a variety of word problems, representing the whole range of problem types listed in Table 11.1, should be used during instruction.

A second, more hypothetical implication that derives from the research findings described in the preceding section is that instruction should develop in children in much more explicit and systematic way than has been done hitherto. Abstract knowledge structures representing the distinct semantic problem types, together with the ability to use these semantic problem schemes in constructing appropriate problem representations should be developed. During the last few years, a number of researchers have implemented this idea in exploratory teaching experiments (see e.g. De Corte and Verschaffel, 1985b; Willis and Fuson, 1988). In these teaching experiments, the child is taught to draw and complete a diagram that matches the semantic structure of the problem, before performing any computation. The main function of these activities is to help children to become aware of the essential semantic relations between the quantities involved in word problems, and to use that knowledge for understanding such problems. Figure 11.1 shows the different diagrams for representing the three main addition and subtraction problem types (change, combine and compare) that were taught to a group of first-graders in our teaching experiment (De Corte and Verschaffel, 1985b).

Our experimental programme for one first-grade class, implemented over 6 months, seemed to have a positive effect on the children's word problem solving. The exploratory nature of this study warns against hasty and definitive conclusions. However, taking into account similar findings obtained by other researchers (see, e.g. Willis and Fuson, 1988), it is our conviction that the principle of explicitly teaching multiple forms of graphic representation deserves further study.

A third practical recommendation that derives from our findings is that writers of textbooks for young children should pay more attention to the appropriate formulation of word problems, and not concentrate only on the purely arithmetic aspects. Our analysis of six Flemish instructional programmes showed that in most current textbooks for elementary mathematics education, word problems are usually stated very briefly, sometimes even ambiguously, at least for someone who is not familiar with the standard problem situations (De Corte *et al.*, 1985a). Suggestions concerning the direction in which one can search for rewordings that can help children to construct an appropriate problem

Fig. 11.1. Diagrams for representing the three main categories of addition and subtraction word problems (De Corte and Verschaffel, 1985b)

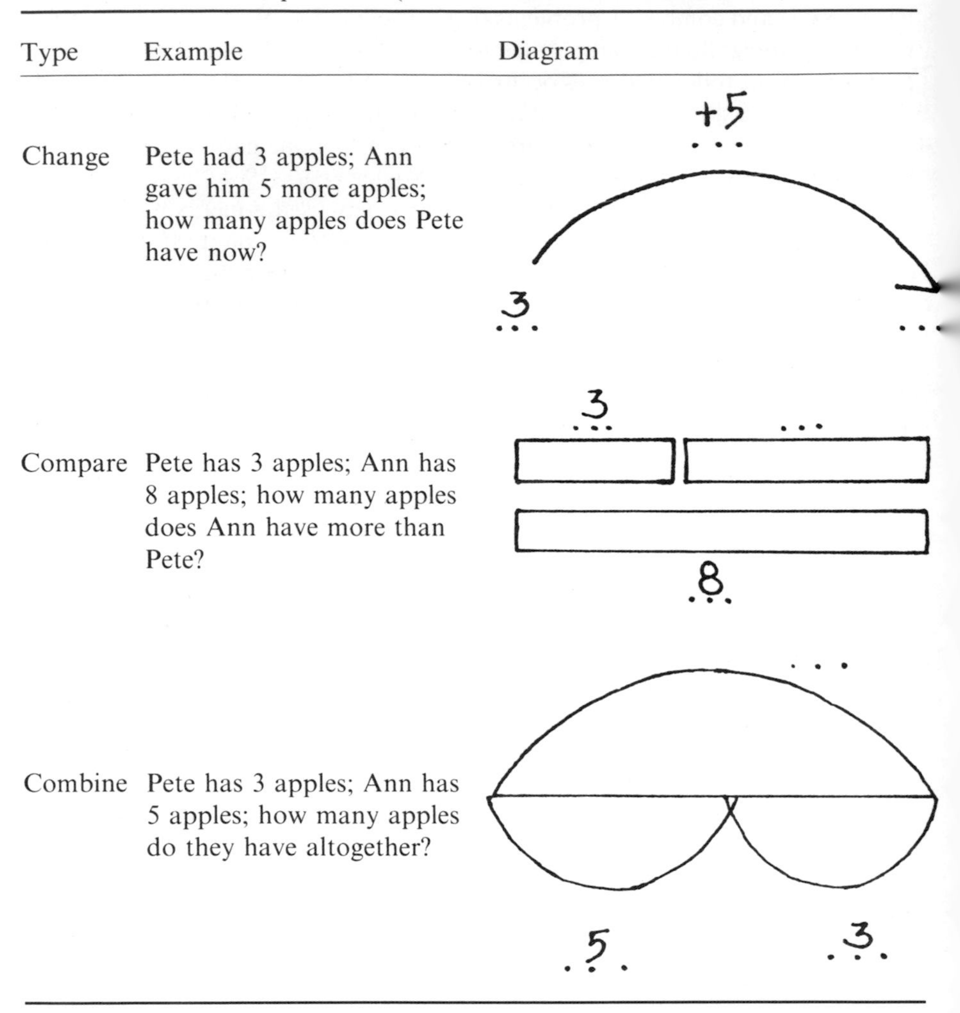

Type	Example	Diagram
Change	Pete had 3 apples; Ann gave him 5 more apples; how many apples does Pete have now?	
Compare	Pete has 3 apples; Ann has 8 apples; how many apples does Ann have more than Pete?	
Combine	Pete has 3 apples; Ann has 5 apples; how many apples do they have altogether?	

representation (and/or find an appropriate solution strategy) in a largely bottom-up way are given on pp. 123–5. At the same time, this can facilitate their understanding of the underlying semantic relations, and as such constitute a starting point for the transition to a top-down approach of verbal problems.

Conclusions

The research findings presented here do not cover the whole area of learning and teaching word problems, not even the restricted domain of simple addition and subtraction problems. Therefore, we conclude with a few more research-based considerations.

First, it has been found that children spontaneously use a wide variety of informal solution strategies to solve word problems (De Corte and Verschaffel, 1987). Taking into account the general principle that instruction should be linked with children's prior knowledge and skills, we strongly argue in favour of matching the teaching of word problems explicitly with children's informal solution strategies. Moreover, the finding that children who have not yet had formal instruction can nevertheless already solve simple word problems using a wide variety of informal strategies that model closely the semantic structure of the distinct problem types, suggests that verbal problems could be attributed a new function in maths education; they could indeed be mobilized in the first grades to promote a thorough understanding of the formal arithmetical concepts and operations, instead of being assigned almost exclusively an application function. In accordance with this view, the experimental teaching experiment referred to (De Corte and Verschaffel, 1985b) does not postpone the teaching of word problems until children have learned the formal operations of addition and subtraction; on the contrary, word problems are presented before introducing these arithmetic operations and the related number sentences.

Secondly, recent research has shown that children's errors on word problems are very often remarkably systematic: they result from misconceptions of the problem situation, which are due to an insufficient mastery of the semantic schemes underlying the problems. For example, some children misinterpreted the sentence 'Pete and Ann have 9 apples altogether' in a combine 2 problem as follows: 'Pete and Ann *both* possess 9 apples' (see De Corte and Verschaffel, 1985a, and Riley *et al.*, 1983, for additional examples). This finding warns us that we should not be too hasty in interpreting errors as being the result of trial-and-error behaviour or sloppiness. Moreover, it underlines that effective instruction and remediation in mathematics requires that teachers have a substantial knowledge and understanding of children's misconceptions and incorrect strategies.

Thirdly, we have criticized some typical school word problems because of their condensed and stereotyped nature, leading young children into difficulties when building an appropriate problem representation (see p. 126). However, the problems being used in present-day instructional practice and in recent investigations, can be criticized for another reason. Word problems can be situated along a 'reality' dimension, ranging from stereotyped, content-lean problems on the one hand, to more attractive, realistic, rich problem situations on the other (De Corte and Verschaffel, 1987; Nesher, 1980; Treffers and Goffree, 1985). An analysis of problems in current arithmetic textbooks and tests reveals that most of them deal with situations that are rather meaningless and artificial from the child's point of view. Therefore, these problems should be supplemented with so-called context problems which refer to situations that are much richer and more closely related to children's real-life experiences than traditional word problems. This does not imply that we argue against the further use of more traditional types of word problems. We would rather advise the concurrent application of a variety of problems along the reality dimension.

Fourthly, the practical suggestions formulated in this chapter have focused on the development of pupils' domain-specific knowledge base, especially the acquisition of semantic schemes. However, this attention to domain-specific knowledge should be complemented with instruction aimed at improving children's strategic knowledge and skills. Indeed, several studies from cognitive psychology suggest that good and weak problem solvers do not only differ with respect to the content of the organization of their domain-specific knowledge base, but also in terms of their ability to apply during problem solving all kinds of cognitive strategies as well as metacognitive knowledge and skills. This research has shown that 'what one knows or believes about oneself is a learner or doer of mathematics and how one controls and regulates one's behaviour while working through mathematical tasks' (Garofalo, 1987: 22), also have powerful effects on one's performance. Consequently, instruction should likewise be explicitly aimed at facilitating the acquisition, application and generalisation of these cognitive and metacognitive skills (Montague and Bos, 1986).

Further reading

Carpenter, T.P. (1985). Learning to add and subtract: An exercise in problem solving. In E. Silver (ed.), *Problem solving: Multiple research perspectives.* Philadelphia: Franklin Institute Press.

De Corte, E. and Verschaffel, L. (1985). Beginning first graders' initial representation of arithmetic word problems. *Journal of Mathematical Behavior*, **4**, 3–21.

De Corte, E. and Verschaffel, L. (1987). The effect of semantic structure on first graders' solution strategies of elementary addition and subtraction word problems. *Journal for Research in Mathematics Education*, **18**, 363–81.

Riley, M.S., Greeno, J.G. and Heller, J.I. (1983). Development of children's problem-solving ability in arithmetic. In H.P. Ginsburg (ed.), *The development of mathematical thinking*. London: Academic Press.

12 Order of mention *vs* order of events as determining factors in additive word problems: A developmental approach

Eva Teubal and Pearla Nesher

When a pupil is presented with a word problem in mathematics, he or she is presented with a number of semantic challenges, as De Corte and Verschaffel have shown in Chapter 11. One particular difficulty is the relationship between the order of mention and the order of events in tackling the problem. Language allows us to describe events in a sequence different to that in which they occurred and word problems provide many instances of this discrepancy, as we will show below. To understand how pupils respond to these challenges calls for a close examination of the tasks they encounter and of the strategies they adopt.

This chapter is concerned with the order of mention *vs* order of events discrepancy. We begin with a discussion of the textual organization of common types of word problems. We outline various possible ways in which different sequences of word orders might impact upon young maths pupils. Among these, we focus on two specific order variables that are relevant to many word problem tasks posed to school children, concerned with order of time sequence and order of numerical data. A review of some of our own research follows, in which we tested children on a variety of word problems, involving change and combine problems. Our results show evidence of developmental changes in the relative importance of different factors, especially order of mention. We discuss some of the practical issues arising in this connection.

Background

Additive word problems have been dealt with extensively in the last decade. There is consensus regarding the degrees of difficulty posed by the relative position of the unknown in addition and subtraction number sentences on the one hand, and the level of difficulty posed by different kinds of text in arithmetic word problems on the other. Problems classified in terms of a semantic analysis yield six different types of 'change' problems and six types of 'compare' problems (see Chapter 11 for an introduction to these terms), depending on the position of the unknown in the 'associated number sentences' (Carpenter *et al.*, 1981). What these authors have in mind when referring to the 'associated number sentence' is a unique arithmetical expression which results from the sequential extraction of the numbers in their order of appearance in the text.

Such a conception of the number sentence associated with arithmetic word problems might be adequate for the very restricted set of problems which do not allow for different sequencing of the data, according to different aspects of analysis. It becomes more problematic if one takes into account the possibility of having different sequencing of data: in terms of order of events in real time; in terms of their order of description in the text; and in terms of their yielding a 'canonical' (i.e. ready for solution) number sentence. Let us take, for example, a word problem that Hiebert (1982) posed:

Problem 1
1 Bill had some marbles.
2 Susan gave him 3 more marbles.
3 Now he has 8 marbles altogether.
4 How many marbles did Bill have to begin with?

In such a problem, the temporal sequence of the data presented by the text matches the order of events in real life and also matches the order of the data in an arithmetic sentence, such as $\_\_ + 3 = 8$ (which is *not* a canonical arithmetic sentence). But consider, however, an alternative formulation of the events described in problem 1:

Problem 2
(3) Bill has now 8 marbles, after
(2) having received from Susan 3 marbles.
(1–4) How many marbles did Bill have to begin with?

Such a problem presents the same information in terms of the needed arithmetic solution but reverses the time order of the events. Without doubt such a formulation is rather awkward, but it serves to contrast the various order sequences we are interested in. The information is presented in the text so that the order of mention of events is the reverse of their order of occurrence. In other words, what is happening now is mentioned first, and what happened earlier is mentioned later. The question which is posed at the end refers to something which occurred at the very beginning (which was also the case in problem 1).

At this point, it is useful to take note of an important distinction between the levels of analysis required by such problems. This is a distinction introduced by Kintsch (1986) between the *text base* and the *situation model*. The text base is the explicit, ordered information presented in the literal description, i.e. what is said or written. The situation model is that which is alluded to by the text, i.e. that which is talked (or written) about. As we have seen, the order of the text need not be the same as the order of the situation to which it relates. Thus, the order of events in the real world and the order of their appearance in the text might present the 'processor' of the information – the child attempting to solve the word problem – with a conflict, when moving from one level of analysis (the base text) to the other (the situation model). The 'associated numerical sentence' corresponding to the second text for our revised problem 2 might now be $8-3=\_\_$, which is a canonical sentence from the arithmetical operation point of view.

If one is interested in children's processing of word problem texts (and their translation into arithmetic sentences) and if the distinction between the text base and the situation model (according to Kintsch) is significant in this process, this cannot be studied when using texts in which the order of mention of events matches their temporal sequence (as presented in problem 1). Furthermore, in a text such as in problem 1, the translation into a non-canonic arithmetic expression is a fairly straightforward move; the burden of the solution process is thus transferred to the solution of a non-canonical arithmetic sentence (i.e. $\_\_ + 3=8$), rather than being upon the translation from the text to the arithmetic expression. On the other hand, in a text such as presented in problem 2, the difficulty lies in constructing the situational model corresponding to the text base and, once this is accomplished, there is no problem in solving the seemingly 'associated' canonical arithmetic expression (i.e. $8-3=\_\_$).

It is important to notice that the two examples above present texts which describe the same events (situational model) but differ in the order of presentation of the numerical data and the order of presentation of the events in the text. In problem 1, the order of mention in the text is as follows: first to be mentioned is the set with an unknown cardinal number (a subset from the logical point of view); secondly, the second subset with its associated cardinal number (3); thirdly, the union set with its associated number (8). Finally, there is a question regarding the subset mentioned at the beginning with the unknown number. In problem 2, on the other hand, the order of numerical mention is: first, the union set with its associated number (8) and, secondly, one of the subsets with its associated number (3). Finally, there is the question concerning the unknown cardinal number of the second subset. Note that this formulation is shorter than in problem 1, in which a reference to the subset with the unknown number is made twice: once in the information component at the beginning and once in the question component at the end.

To summarize, our analysis leads us to propose that three different aspects of the text of a word problem are relevant to understanding the steps involved in its solution:

1 The logico-mathematical aspect (i.e. whether it calls for an addition or subtraction operation, etc).
2 The 'real-life' situation described by the text.
3 The order of mention in the text of:
   (i) the events,
   (ii) the numerical data.

We turn next to an examination of how these factors bear on children's performances.

Order of events and order of numerical data as variables in children's word problem solutions

In view of the above considerations, we have conducted research designed to allow comparison between the impact of the different aspects of sequencing at various age levels. The novelty of this study is the attempt to contrast two different *order* variables. One of them – the order of events in terms of actual time sequence – concerns the child's everyday experience and represents his or her understanding of the coherence of the situational model (Kintsch, 1986). The second order variable – order of numerical data in the text – concerns the child's newly acquired formal knowledge: the knowledge of arithmetic, its operations, as well as its constraints. The impact of these two orders was examined at two different ages in order to reveal a developmental trend.

The tasks

The tasks used were six types of verbal arithmetic problems, four of which were of similar structure (change problems) which are typical of time sequence, and two were of a different semantic structure (combine problems) in which the time order had no relevance. The word problems were composed according to a manipulation of variables derived from the three aspects of analysis mentioned above:

1 *The logico-mathematical aspect.* The manipulation of this variable yielded values concerning the presence or absence of numbers corresponding to the three possible sets involved in any additive situation. Within this variable, we manipulated two possible values: (1) the presence of the numerical information corresponding to the two subsets and the absence of the number corresponding to their union (this leads to an addition sentence if solved canonically); and (2) the presence of the numbers corresponding to the union set and one of the subsets, and the absence of the number corresponding to the other subset (this yields a subtraction sentence if solved canonically).
2 *The model situation aspect described by the text.* Two values of this variable were considered here: combine problems and change problems.
3 *The order of mention in the text aspect* is in fact two distinct variables to which we will refer separately:
   (a) *The order of mention in the text of the sequence of events.* Two values of this variable were considered here:
   (i) *The ordinary order*: the event mentioned first at time 1 (T_1), then the event mentioned second at time 2 (T_2) and, finally, the event mentioned at a later time, i.e. time 3 (T_3). In short, the order is T_1, T_2, T_3.
   (ii) The fully reversed order of (i), i.e. T_3, T_2, T_1.

(b) *The order of mention in the text of the numerical data needed for solution.* Let S represent the quantity corresponding to a subset, and U the quantity corresponding to the union set; then, there are few orders of the three basic components of every additive problem (S-S-U). But there is also the consideration as to which one has the unknown number, and there are more combinations of these. In the present experiment, we dealt with the following values of this order variable:

S S Ⓤ
U S Ⓢ
S Ⓢ U

All of the possible combinations of the above-mentioned values of variables yielded a pool of possible versions of story problems. It was decided to include those versions which allowed us to contrast the relative influence of the order of occurrence of events vs their order of mention in the text and the order of the numerical data.

The types of word problems presented in terms of the above-mentioned variables used in the study are listed in Table 12.1. The problems were assigned so

Table 12.1 Change problems

Problem number	*Problem structure*	*Problem text (example)*
1	S Ⓢ U T_1 T_2 T_3	At school in the morning there were 20 pupils inside the building. More pupils entered the building later in the day, at break-time. There are now 80 pupils at school. How many pupils entered the building at break-time?
2	U S Ⓢ T_1 T_2 T_3	There were 80 pupils in the auditorium, 20 pupils went our during break-time. How many pupils were left in the auditorium?
3	U S Ⓢ T_3 T_2 T_1	Now there are 80 pupils in the auditorium, after the entrance of 20 pupils during the break. How many pupils were there at school before break-time?
4	S S Ⓤ T_3 T_2 T_1	Now there are 20 pupils in the auditorium, after the exit of 80 pupils during break-time. How many pupils were there in the auditorium before break-time?
Combine problems		
5	S Ⓢ U	There are 20 boys in the class. There are girls as well. If the total number of pupils is 80, how many girls are there in the class?
6	U S Ⓢ	There are 80 pupils in the class; 20 of them are girls. How many boys are there in the class?

Note: Each of the above six structures appeared in six different story contexts.
○ marks the unknown number.

that each pupil solved six problems, each one belonging to a different type of structure and a different 'story context'. Examining the question of 'change' *vs* 'combine' problems was not the purpose of this study. The reason for including the combine problems was that they would allow for a comparison between those texts in which there was (supposedly) an interaction between the order of mention of events and the position of the unknown *vs* those texts in which the only factor (supposedly) influencing the information processing is the position of the unknown. The hypothesis concerning the change and combine types of texts was that 'combine' problems (in which order of events does not apply) would represent a task whose degree of difficulty should be somewhere between a change problem of the order T_1, T_2, T_3 and a change problem of the order T_3, T_2, T_1.

The tasks were presented to 30 children aged 7–8 years (second grade) and 30 children aged 12–13 years (sixth grade). The children were interviewed individually and their replies provided the basis for the data analysis. We will concentrate on two aspects only of performance on our tasks: (1) the percentage of success for each problem story, and (2) the strategy chosen to solve each problem story. We were interested in the percentage of success only in conjunction with the solution path (i.e. the choice of strategy and algorithm). In the solution path, the following strategies were considered as representing different modes of information processing:

1 *Sequential solution.* The children wrote down arithmetical sentences in which the data appeared in the same order as in the text. This resulted in certain problems in a canonical sentence and in others in a non-canonical sentence.
2 *Reorganization of the text.* The children wrote down arithmetical sentences in which the numerical data appeared in an order different from those in the text. Reorganization of the data could respond to two different needs: either to arrive at a canonical sentence, or to arrive at a data sequence which reflected the natural sequence of events in the world, i.e. the number associated with the earliest event (T_1) appears first, which is followed by the number associated with the second event (T_2), and the third number in the arithmetic sentence is associated with the third and last event (T_3). Note that such a reorganization does not ensure a canonical sentence.

These particular problem texts were selected because they lent themselves to different types of organization. They corresponded to the different kinds of analysis whose relative influence we were interested in comparing.

Table 12.2 presents alternative ways of organizing the problem data, so that maths sentences of different forms are obtained, depending upon the relative weight of the order variables in the process of being solved. Table 12.3 presents the percentages of correct answers for each question, and the percentages of solution strategies. As can be seen in Table 12.3, the 'percentage correct' can serve as an index of the degree of difficulty posed by this kind of problem for second-graders. As far as sixth-graders are concerned, there is not enough variability in this measure to provide an indicator of the degrees of difficulty;

Table 12.2 Possible mathematical sentences leading to a correct solution

	Sentence resulting from order of mention in the text	*Sentence resulting from reorganization of data*
1	S + ○ = U	U − S = ○*
	T1 T2 T3	T3 T1 T2
2	U − S = ○*	○ + S = U
	T1 T2 T3	T3 T2 T1
3	U − S = ○*	○ + S = U
	T3 T2 T1	T1 T2 T3
4	S + S = ○*	○ − S = S
	T3 T2 T1	T1 T2 T3
5	S + ○ = U	U − S = ○*
6	U − S = ○*	S + ○ = U

* Canonic expression; ○ marks the unknown number.

Table 12.3 Solution strategies adopted by 2nd and 6th grade subjects

	Percentage correct		*Percentage of sequential ordering according to mention*		*Percentage of reorganization of data*	
Problem number	*2nd grade*	*6th grade*	*2nd grade*	*6th grade*	*2nd grade*	*6th grade*
1	90	95	100	30	0	70
2	93	100	94	97	6	3
3	68	97	77	84	23	16
4	68	85	71	90	36	10
5	93	94	87	17	13	83
6	87	100	52	100	48	0

Note: Sixth grade, $N=45$; second grade, $N=45$.

reaction time might be a more interesting index at this level. Therefore, we will take a look at the degree of difficulty posed by the questions for second-graders, and at the solution paths characteristic of different structures at different age levels.

Solution strategies

As to level of difficulty, it is clear that the easiest problems are the canonical subtraction change problem (problem 2) and the non-canonical 'addition' combine problem (problem 5). An examination of the easiest change problems (problems 1 and 2) indicates that the order of numerical data in them is S-S-U, and the order of events is T_1, T_2, T_3; whereas the most difficult problems are those in which the (real) time sequence is violated in the story. The events are presented in the reverse order to that of their occurrence (i.e. T_3, T_2, T_1), while the logical order of S-S-U is maintained (problems 3 and 4).

When looking at the solution path followed when solving the various types of problems, one notes that younger subjects are influenced much more by the order of the text. Whenever the information in change problems is reorganized, it is done so that it corresponds with the order of events in real time. When it is done with combine problems, it is done so as to transform a subtraction exercise into an addition exercise with a 'missing addend addition' rather than the canonical subtraction which is the direct solution of problem 6. Problem 6 is the one in which the highest proportion of 'reorganizers' among second-graders is found. This is consistent with Piaget's contention that logical subtraction, which is at the heart of class-inclusion problems, is still hard at this age (Piaget, 1952). It can be seen from Table 12.3 that there was not one single sixth-grader who 'reorganized' problem 6 in his or her solution.

Table 12.3 shows that older subjects are less influenced by the order of the text. The dominant factor in their choice of solution path appears to be 'reaching a canonical solution' (the unknown number appearing at the end of the arithmetic sentence). This means the dominant influence is that of the logico-mathematical level of the text.

Practical implications

Although we have directly tested means of interviewing, we can point to two simple but important practical considerations that merit attention.

1 Like De Corte and Verschaffel, we conclude that teachers and textbook writers should attend usefully to the relationship between the order of mention and the order of arithmetical events stipulated in any word problem. In particular, when these conflict, there is potential for confusion or misunderstanding, especially with younger pupils.
2 The developmental shift to which we have pointed may mean that different strategies may be appropriate for different age groups. For example, it may be best to minimize or avoid a conflict between order of mention and order of event for younger pupils and to concentrate on encouraging their mastery of the arithmetical steps. However, the relationship between language and mathematics is a complex one, and the conflicts we have been discussing here must be tackled at some stage. Hence, it may be desirable to guide older pupils to an understanding of the varieties of relationship that can hold between the text base and the situation model.

Conclusions

The main question we have addressed in this chapter has concerned the relative influence of two types of order of the information in the text: the first is the order of mention of the events, and the second is the order of the numerical data accompanying the different sets described in the text. The first order assists in constructing the situation model and arriving at the reference, i.e. at understanding in a coherent manner the sequence of events and actions described in the text. The second order assists in the formal solution of the task, i.e. finding a missing number by arithmetic calculation. We have conjectured that these two different factors do not necessarily develop in parallel. While the function of reference of natural language is developed from the start of language usage, the calculation function starts mostly with formal schooling. These factors, therefore, may affect children's performances differently at different ages. In our experiment, the order of mention of events in the text compared with their order of occurrence, appears to be the dominant factor which affects information processing in young children. The direction of the activity as described in real time (increase or decrease) and its correspondence to a canonical addition or subtraction, is the easiest case and first learned. When such a correspondence is violated and two orders in conflicting directions are presented simultaneously to the child, a further elaboration of the text appears to be needed to gain agreement between the conflicting pieces of information. The present study suggests that young children resolve this conflict differently from older children. Young children rearrange the text so as to arrive at the proper time sequence of the events at the price of a non-canonical arithmetic sentence, whereas older children disregard the order of mention of the events and they are able to appreciate a canonic arithmetic sentence more which is amenable to direct solution.

Further reading

Kintsch, W. (1986). Learning from text. *Cognition and Instruction*, **3**(2), 87–108.

Nesher, P. and Katriel, T. (1978). Two cognitive modes in arithmetic word problem solving. In E. Cohors-Fresenbory and I. Wachsmuth (eds), *Proceedings of the 2nd International Conference of the Group for the Psychology of Mathematics Education*, pp. 226–41, Osnabruck.

Nesher, P. and Teubal, E. (1975). Verbal cues as an interfering factor in verbal problem solving. *Educational Studies in Mathematics*, **6**, 41–51.

Section 5

Discussion

13 Peer discussion in the context of mathematical problem solving

Susan Pirie

Languaging is not in the brain – it is among us. Knowledge is not in the brain – it is in the interaction (Tomm, 1989).

This chapter is concerned with some of the characteristics of pupils' discussions in mathematical lessons. Discussion is defined here to be purposeful talk on a mathematical subject in which there are genuine pupil contributions and interaction. Peer discussion of this kind (as distinguished from teacher–pupil exchanges) has received scant attention in the past, but it is important as an arena for learning which fills an ever increasing portion of lesson time due to the growing encouragement of cooperative group work in schools (see Hoyles *et al.*, this volume). This chapter considers how thinking and problem-solving abilities are influenced by, and themselves in turn influence, the language used by a small group of secondary school pupils working together on a mathematical problem.

The chapter begins with a brief overview of some of the issues involved in studying discussion. Then a sample discussion is presented and analysed. The analyses are concerned with features of pupil discussion which may be found in many other peer interactions in the mathematics classroom, and I hope the account will provide readers with guidelines for studying discussions which they themselves have the opportunity to observe or record. The analyses focus in turn on the language of problem solving, the influence of language on the task, and pupils' interpretations of the problem statement that they were set to work on. Some practical implications of mathematical discussion are considered.

Studying discussion

A transcript of a verbal interaction allows us to stop time, to wind back the clock and repeat transient utterings, and therefore enables us to form considered hypotheses as to the thought processes of the participants at that moment. Such hypothesizing, if it is to have any value, is no easy task. It is all too simple to say 'Obviously what they mean is . . .', yet to gain true insights 'one has to work rather hard to make the effort of will and imagination to render what is familiar, strange' (Atkinson, 1981).

A variety of aspects can act as the foci for analysis: the roles adopted by the participants and their relative degrees of participation; the effective components of their interactions; their personal concerns and interpretations; the mathematical processes and strategies they bring to bear on the problem; and many more. Here, however, I wish to look at the language of problem solving: at what this language is; at how languaging influences doing; and whether verbalization constrains thinking. In so doing, account must be taken of external influences such as the past experiences of the participants, both personal and mathematical, the broad societal expectations concerning the conducting of discussions, the more specific expectations of discussion within the context of the school, in particular the mathematical classroom, and the covert 'rules' associated with behaviour in group work. Space does not permit a detailed review of how each of these is relevant to the discussion we will consider here, but the reader might find it interesting to explore the transcript with a different focus in mind (and see Pirie, 1988 and Pirie *et al.*, 1989, for further illustrations, as well as Edwards and Mercer, 1987, for an account of the construction of common knowledge in the classroom).

A sample discussion and analysis

The transcript to be examined is presented with only one section cut from the original discussion (see D below), although names have been altered. The problem was presented in written form with the words 'Here, have a go at this.' The data have been taken from a project investigating the effects of discussion on mathematical understanding (Pirie, 1988; Pirie and Schwarzenberger, 1988). The three participants, all girls, had no expectation of having to write the problem up either during or after the session.

A

Claire [reading questions]	**Ann**	**Laura**
		[Takes a pencil and paper]
74 is closer to 81 than 64, so the square root of 74 is closer to 9 than 8. True? Generalizable?		
Proof?	What?	What's the middle word?
Gener izable		Generalizable in other words
Generalizable		Yes

74 is closer to 81 than 64	Yes	Yes
OK	Yes it's true	
		Yes I agree with that
There we'd better try it out		
	What's the square root of 74? [gets calculator]	
Write the square numbers down the edge of the page		
		There's only 3
Write the square number down the edge of the page		
		What square number?
Well, like 81 and 64 are square numbers.		
So write . . . 4		81 . . . [writes 81]
No, no in order Laura	'm	
Write 4	That bit's true	What is the order?
1 'Cos that's the square root of a square, 4, 9, 16, 25		
		Why?
	The square root of 74	
81 and 64 . . . 74 is closer to the square 9^2 than it is to 64^2		
	Shall I tell you what the square root of it is? 8.6023253	
8.6 yes it's closer to 9. Closer to 9		
	So it doesn't have to be square number like that	
2 No look. These two are square numbers. And you take a number that's not a square number and it's closer to which of the two square numbers it's between and then you'll know its square root is closer to the square root of those square numbers. No Ann don't		Wonderful
	Yes	

3 So write 2, 4. I mean 4, 9, 16, 25, 36, 49		
		[writes 4, 9, 16, 25, 36, 49, 56 in a vertical column]
	56	
		64 actually [changes 56 to 64]
er yes Oh yes		
	Now what happens?	
81, 100		[writes 81,100 in column]
	That'll do. OK	
Now one, choose one in between those two		
	In between which two?	
4 and 9		
	5. Have that one top	
Which is it closer to?		
		4
	Ah no	
So now if you pick a number, pick a number, pick a number	5, 5 The square root of 5 . . .	5, we already have
. . . is going to be closer to 2 than 3		
	It's going to be 2.236068. So yes, that works	
Pick another number		
	Right. Which is it going to be closer? What's it going to be closer to?	
		4, let's see . . . 2
	2.4 yes	
B **4** No hang on a moment. We'll have to try a middle . . .		**4** Try the middle one between them
No wait a minute. Yes that's what I want to know. Something exactly half way between. So what's it got to be? It's got to be between an odd number and an even		

5 number, hasn't it, to be half way between?		
		6 Well do it in between 9 and 16 then
K.O		
7 Isn't that half way in between ? [points to 6]		
		No I shouldn't think so Claire
8 5 and 8, 6 and 7, no		
	9 6 and 7 that's true	
		10 6 and 7? 7's between 9 and 16. Which is 3, 3 and 1. So what's 4 and 9?
	13	
		11 13. 13 is the middle of 9 and 16. OK?
11 I was wondering how you did it		
		Because 7 is 3 and 3 on each side and 1 extra in the middle. Yes, OK?
. . . and that 1 there.		
Well! Fine, yes, OK	Yes	
7		
		7 So it's going to be . . .
	Too far	
Let's see what it's going to be . . .		
Well do that one then.	No, no, no that one's just easier	
Um . . . 9	What? What do you want me to do?	
		4 and 9 . . . 13 square root it
13 square root of it. Hang on a minute and I'll tell you what it's going to be. 3 and 4. It's going to be 3.5		
	3.6	
		That's very interesting

No Ann, no It's um OK except for half way cases	I'm afraid . . . that is	
		And then it does what it likes
Perhaps it always goes up Perhaps it always goes up		
	What is half of 4 and 9?	
P'raps it always goes up		
		4 and 9 is 5, 2.5
	No *that's* right	
		Try 6.5
	Excuse me, that's right	
		No it's not
	'Cos 13 is nearer 6 than 9	
		No it's not
	Yes it is	
		6 and 9? We're not even doing between 6 and 9
		No it's not. It's exactly in the middle
1, 2, 3, 4, 5, 6, 7, 8, 9, 10, 11, 12, 13, 14	No it's not 'cos 9+4 is 13. 16−3 is 13, therefore it is closer to 16 than 9	
		Mm
	You can't have in between cases	
		12 Of course you can
	No you can't unless you have something point five; 12.5	
12 There must be a way you can have in between numbers		
		12 No Ann there must be in between cases. Don't be silly
	Not whole numbers	
Yes		Yes
	No . . .	

If you have 3. If you have 1 and 3 the midway is 2		
	Yes, but they're not square numbers. We want square numbers	
There must be some!		
		So? There is no reason why there shouldn't be among the square numbers
	12.5 That's an in between isn't it?	
		Yes. Of what?
	13 Yes, so that's 3.5. It should be . . . That's over – I've rounded it [using calculator]	
But there must be a half way one		
	There isn't because they're all even and odd if you see what I mean	
Mmm		
	and you can't get half way . . .	
		14 So it's all right for one decimal place
	Yes	
15 But normally you have things to 3 significant figures		
	16 It doesn't work to 3 sig. fig.	
		17 So it only tells you roughly . . .

C		
18 We have to find a proof So we need to call it x . . .		
	So it's true. It's gen . . . gener izable	
. . . generalizable . . .		
	Yes, knew I'd got something wrong	

	You can apply it to anything rather than just that			
		Yes I know what you mean I just can't say it		
	OK Umm so we have a number. Er can you write the square number each side?			
				Mm What!
	So if you have a number x and that's your number like 74, how can you write the square number of each side of it?			
				Two consecutive square numbers
		Uh?		
	No, I want to be able to write this in algebra so I can prove it			
		Square root of x		
	I don't know how . . .			
				I know what you mean
		Um		
	Yes, the square root of x plus . . .			
				What?
		Hang on You have to have it squared		
				Sorry?
	You have to find, to find the two square numbers each side of it	x^2 and y^2		
		You have to have it as x^2 and y^2		
19	x^2 and y^2 are two consecutive numbers			
20	and z comes in between			
			20	But they're not consecutive numbers are they? They're square numbers

Two consecutive square numbers and z		
	Yes No that's not what consecutive numbers mean you know	
		x and y. Consecutive 6 and 7. Consecutive numbers
	That's not what consecutive square numbers means. There's no such thing as consecutive square numbers	
There is now		
		You've got to do it. They're two consecutive numbers, they're two consecutive numbers
Um . . . say you have x^2		x and y are consecutive numbers
	What do you mean?	
. . . x and y are consecutive numbers?		
	Yes, like 8 and 9	
		Yes
OK. I'll allow you that much.		
	You can't have consecutive square numbers	
All right, consecutive numbers squared. So x and y are consecutive numbers and . . . they're consecutive numbers and *then* you square them		Squared
	OK	
		20 z comes between [writes x^2 z y^2]
21 and x^2 . . . is small. I know, I know, x^2 is smaller than z which is smaller than y^2, and x . . .		

		I think it should be z^2 Claire not . . . [Adds $<$ $<$ to above]
x, how can you write the numbers		
		I think it should be z^2 [changes z to z^2]
	No because it's not a square number	
Oh we can root it, Laura, we can root it. It's not an exact thing squared		'tis, we can square it
	So x is smaller	
So, no hang on a minute. I want to write that x and y are consecutive		
		$x=y+1$
	No that doesn't work because x is smaller than y so it's $y-1$	
$y=x-1$ and that and now we are going to say	Yes	[writes $y=x-1$]
	x is smaller than square root of z which is smaller than y	[writes $x<z<y$]
and now you want to say that z [snatches paper]		
'if x^2 is less than z^2' no we're not having z^2 because it's not a square	Yes	
of an exact number. Put a z and we'll square root it. I'll put a square root sign . . .	No because	[snatches paper back]
[takes paper, adds $\surd$ to z in $x<z<y$]		It doesn't make any difference does it? OK so put it on
Yes it does, so I've written it the other way		
	[paper snatched back and forth]	. . . I've put it down [writes $x^2<z<y^2$]

It doesn't matter. I want to read it Laura '$x^2<z<y^2$' [gives paper back]		OK
So, umm . . .	That's right	
		. . . Yes, but umm you helped
	Yes it does	
Um, so now how do you write that the square root of z. . . . Then you can write the square root of z is going to be		x is square
	If x is smaller than the square root of z which is smaller than y . . .	
Uh!		
I just don't, yes I don't know . . .		
I'm asking . . .		
OK, write nearer than, nearer to whichever one it's nearer to . . .		**22** How are you going to write 'nearer than'? How are you going to write 'nearer than'?
I know, you write . . . $z-x^2$		Hold on [writes $z-x^2$]
$z-x^2$	**22** Is less than . . .	
is equal um . . . if $z-x^2$ is less than y^2-z then um . . .		. . . y^2-z [writes $<y^2-z$]
	What?	
23 Are you lost?		
	Yes, I am for the moment.	No then . . .
then $z-x$ $z-x$ just finish it first then I'll explain it to you		
	OK	
		OK [writes z] No then z'll be nearer
then $z-x$		

then $z-x$		OK. Yes, then square root of z will be nearer
minus x will be less than $y-z$ and vice-versa		Minus x [writes $-x$] y−square root of z [writes $<y-z$] and if . . . [writes if. Goes back and puts $\surd$ on first z]
Um you know . . . you understand [to Ann] $y=x+1$		
	Yes	
'cos they're consecutive numbers		
You understand that,	Yes, yes I got that bit. It's just there I get . . .	It's irrelevant anyway
Well um then if umm $z-x^2$ is less than y^2-z that means it's closer to the x^2		
	Oh right, OK	
Then the square root of z is closer to the x		
	Yes, Yes	
$z-x$ is less than		
	Yes	
Have a go. Now try it with a number		
D I was having . . .		
24 No you have to pick your z, 'cos z is the number we are doing it do		No you
Yes you do. It's a way of using it if you have a number and you want to know roughly what its square, what its square root is going to be . . . so pick 92 as z		
25 Write, $z=92$. Let z equal 92. check, OK		**25** Let $z=92$ [writes $z=92$]

	25 'Let', put 'let'	
26 No 'cos we had it either way. We had an 'if'		
	In that case you've got your x and y the wrong way round	
27 No she hasn't, 'cos you see then what happens Ann is you don't know which of these two numbers z is closer to. You don't know which of them it's closer to and that's what you're finding out and that's why it's wrong to put in your sign at the beginning because you don't know yet. So you work out what each side of the equation is and then you know which way you can have your sign. And which it's closer to and so which its square root is closer to		
E		
So now you can write that um . . . therefore the square um . . . [writes $y-z<z-x$] y minus the square root of z is less than the square root of z minus x		
		Yes . . .
No 'cos now we, now we check it. Making sure.		
So that means the square root of 92	[uses calculator to square root 92]	
is closer to um, Is closer to . . .	to that	
	Hang on	9 than 10 10 than 9

	or that	
y minus it is less than . . . 10 minus it is less than it minus 9. I think 10, it's nearer to 10 than 9		Uhu
	Yes 9.59	
		Yes Good
it works		Good

Before examining the discussion in detail, it is helpful to parse the transcript into five episodes. These are labelled **A** to **E** and indicated by horizontal lines in the transcript

A The girls explore the meaning of the problem statement, by attending to its language, its specific accuracy and general possible truth.

B They attempt to come to grips with the truth of the statement and are led to an apparent contradiction which they are able to resolve for their purposes.

C They move on to trying to generalize using algebraic representation and this results in some close discussion of meaning and symbolic writing.

D Here follows a lengthy, heated, at times acrimonious, argument about writing mathematics when they come to test out their generalized statement. From considerations of space, this section has been omitted except for brief passages.

E Agreement is reached on a concluding action and all three are satisfied that they have solved the given problem.

On a first superficial reading of the transcript, one might be tempted to conclude that they become diverted from the task into a circular exploration of algebraic representation followed by a fruitless argument over how calculations are written down, that they dismiss or ignore incompatibilities within their working and that they fail to answer the questions within the problem statement. Closer analysis of the discourse and its language, however, reveals the problem as *they* saw it and on which they worked. It is this very tension between spoken and symbolic written mathematics which is at the heart of this chapter, which allows us to glimpse some of the mathematical thinking of the participants and which may reveal the effect of verbalization on 'doing'. The problem-solving language used will be closely examined for what it can reveal of, and what effect it has on, the problem-solving process.

The language of problem solving

What then is the language of the problem solver? How do pupils, working cooperatively on a mathematical problem, discuss their task and their thoughts? Even with no teacher present, as in this transcript, the participants *are* aware of formal language rules within the context of the work they are doing. Claire (**25**) moves from saying 'so pick 92 as z' to the formal 'let z equal 92' and this expression 'let' is taken up verbally by both Laura and Ann although not written down. Later, (**26**) Claire also says 'We had an "if"', implying that this word signals a particular type of formal statement. Halliday (1978) develops the notion of a 'register' as 'a set of meanings that is appropriate to a particular function of language, together with the words and structures which express these meanings'. Thus the 'register' used is determined by what a speaker does (as opposed to a 'dialect' which is determined by who the speaker is), and 'is the meaning potential that is accessible in a given social context'. Pimm (1987) explores this idea in depth in terms of the mathematics register and claims that 'access to particular registers, and also awareness of which elements and constructions belong to which, can be extremely important'.

For student mathematics problem solvers, however, it is not clear in which register they ought to be functioning nor in which register they can most effectively function. Are classroom conventions relevant? Is 'teacher–pupil' speak still appropriate? How far can participants, using the same language, be expected to be describing the same mathematical concepts? Is normal social talk apposite? For the novice, there is also the danger that there may be, as yet unrevealed, linguistic rules governing the problem-solving situation. Pupils do not ask these questions, but the success of problem solving may depend crucially upon pupils' perceptions of the unspoken answers, because the language of group problem solving cannot be the language of 'doing' mathematics, since this is symbolic and written, nor solely the register of teaching mathematics, although this can be useful in discussion, nor yet totally the register of social conversation, although the pupils are conversing with their peers.

As the transcript illustrates, the talk moves between technically different registers and this oscillation results in particular behaviours, such as defining some terms and not others. For these pupils 'square' and 'square root' have become 'domesticated' (Halliday, 1978), and therefore open to the social rules of 'best possible' interpretation which allow the ignoring of speech errors (**1**). When qualified by 'consecutive', however, this is no longer the case (**19**), and a precise definition is sought. What is more important, however, is lack of overt awareness of the register in use at any one moment, since this can lead to serious misunderstandings.

The word 'between' causes just such a problem for these pupils. At the start of episode **B**, Laura refers to the 'middle one between' (**4**), Claire to 'half way between' (**5**), Laura to 'in between' (**6**), Claire to 'half way in between' (**7**). The reader is led, as the pupils were, to believe that they are all referring to some central number which could be imagined on the number line from one square number to the next. Claire counts inwards in pairs, '5 and 8, 6 and 7' (**8**) to confirm that 6 (used earlier in the calculation) is not the middle number. Ann, however, sees that '6 *and* 7' (**9**) are the numbers half way between 4 and 9. When Laura states that '7's between 9 and 16', she is using 'between' to mean 'the difference between'. She has jumped from the domesticated, quasi-mathematical meaning of 'between' to its specific and different meaning within the mathematics register. It is true that $16-9=7$, but there are only 6 numbers 10, 11, 12, 13, 14 and 15 'in between' 9 and 16. Unfortunately, she is unaware of her misdistinction and proceeds to split 7 into $3+1+3$ and suggests that $9+3+1=13$ is the middle number they seek. It is this simple switch between accepted registers which may have been responsible for the considerable, confused discussion which follows and in which Ann appears to be unable to convince Claire and Laura of the true nature of 'an in between'. Later in the transcript, Claire speaks of x and y and the 'z comes in between' (**20**), and Laura picks this up in its physical positional meaning and writes $x<z<y$ before moving to its mathematical meaning which can be conveyed by x^2zy^2.

There is no easy solution to difficulties of this type because inherent in the nature of problem solving is a need to move back and forth between fields of concern (Pirie and Kieren, 1989). This phenomenon is even more pronounced in the solution of problems that have their roots in real-life situations. A deeper difficulty lies, however, in the notion of a mathematics register. 'Language, unlike mathematics, is not clearcut or precise. It is a natural human creation and . . . is inherently messy' (Halliday, 1978). The boundaries between registers cannot be defined. Words and structures are borrowed from one register and redefined in another. Technical terms may be created, but then with use gradually accrue broader meanings until they acquire meanings that are 'in the air' even for those who are unfamiliar with the original technical terms. One of the problems when trying to assess pupils' mathematical understandings is the need to decide the pupils' meanings for the words they use (see chapters by Durkin and Shire and by Nolder, this volume). Spoken language is a growing and evolving phenomenon, but the language of mathematics is written, symbolic and virtually static. This leads inevitably to changing or multiple relationships between the spoken words and the written symbols. Wrong assumptions of pupils' meanings can lead to unproductive teaching or, worse, misinformation by the pupils of their concept links.

The influence of language on the task

Reference has been made to the influence of external forces on the language of problem solving but it is also relevant, from the point of view of problem solution, however, to consider the effects of the language used on the task itself. We need to return to the question of whether the course and structure of the cognitive processes might be changed by the verbalization. It is worth keeping in mind some of the findings of 'think aloud' research methods, as they may also apply to group discussion while problem solving. Werner and Kaplan (1963), for example, have shown that verbalization for one's personal future use will be condensed and idiomatic, whereas to communicate information to a stranger, care and thought has to be given to find communal, transpersonal referents. In which category do pupils see their problem-solving peers – as extensions of themselves with shared, idiosyncratic understandings, or as outsiders for whom they have to process their verbalization in extra ways? The instruction 'think aloud' has not here been overtly stated to the pupils but, as has been mentioned, the format of 'working as a group' imposes its unspoken rules and leads to pupils attempting to speak their thinking.

A clear example of one pupil's idiosyncratic thinking aloud that needs to be more extensively explained is seen in the exchange between Laura and Claire (**10**, **11**) over how to find the middle number between 9 and 16. While it is impossible to know how the pupils would have fared individually with the problem if left to solve it in isolation, it seems reasonable to conjecture that, without the verbal confusion caused by the word 'between', the obsession with whole numbers by Laura and Claire, and their 'closing' statements (**17**, **18**), Ann might have gone on to examine in greater depth the 'untruth' of the problem statement. Later, when working on algebraic representation, Claire demonstrates (**23**) that she is well aware both that her train of thought might be lost if she pauses to elaborate for Ann, and that equally such elaboration is necessary for the group to function.

Ericsson and Simon (1980) offer evidence that, when working under a high cognitive load, subjects are apt to stop verbalizing or verbalize incompletely. In the transcript under consideration, however, it can be seen that the covert rules of group work seem to be more powerful than personal inclination: there are very few pauses and the only significant period of silence occurs at the end of **D** and is probably caused by emotions rather than a need to think mathematically. This very lack of quiet, thoughtful moments may corroborate Ericsson's (1975a, b), findings that 'think aloud' subjects tend to plan their moves or think ahead less than the silent control groups. Certainly, the time taken by the pupils on the generalizability of their ideas (episode **C**) illustrates clearly that 'telling what to do' takes precedence over silent strategic planning.

The constraints placed on verbal problem solving by the language of mathematics itself have not yet been examined. It is the very nature of mathematics to encourage through its symbolic, written language the reduction of explained processes and of overt reconstruction of precepts. The power of the subject lies in the ability to move forward without recourse to explanations and to automate intermediate stages of process. In the data processing model proposed by Ericsson and Simon (1980), 'verbal explanations of automated activities' are seen as 'cumbersome and would change the course of the processing from a largely perceptual to a more cognitive one'. Such a verbalized cognitive process could force the speaker to 'generate [the] concepts and principles [needed] from whatever information is currently available to them. The reformulation may not at all reflect the way in which the learning was acually encoded.'

Thus pupils verbalizing mathematical thinking may be found apparently to form their possibly erroneous mathematical beliefs by the force of their own explanatory words. Claire and Laura appear to do this (**12**) when they talk themselves into a position of conviction that integer half way numbers must exist. Both are quite capable of the trivial arithmetic needed to demonstrate the fallacy of their position and even Ann who sees the correct solution does not verbalize how she knows. An alternative effect, suggested by Duncker (1945), is that certain cognitive activities are so automated that problem solvers will not mention them because they are unaware that they have done anything. This can, of course, cause difficulties when the participants do not have the same automated repertoire.

One striking feature of this problem-solving incident is the very little writing down that takes place. Did the group feel inhibited from writing by the implicit group work rule demanding talk? Would Claire have managed to formulate the algebraic expression faster and more easily if she were not also struggling to verbalize it for someone else to write down? Was this apparently massive verbal task responsible for diverting their cognitive attention from the main problem? If one were in a position to answer these questions, and if the answer to any of them were to be 'yes', then serious consideration would need to be paid to the possibility that under some condition the discussion of mathematical working is not conducive to effective problem solving.

The pupils' interpretation of the problem statement

From their closing remarks, it is evident that the pupils believed that they had successfully completed their task. The reader may think otherwise. A scrutiny of the discourse reveals that the problem becomes transformed by the interpretation and reformulation of the language of the original question. As Simon (1979) showed, 'relatively "innocent" changes in the way in which the problem is stated may have major effects on the way in which people encode it'. Very early in the discussion (**2**), Claire re-words the problem in verbally generalized terms which presuppose the *general* 'truth', and this probably sows the seeds for the later qualification of the accuracy of the 'truth' (**14–17**), rather than the exploration of the area within which the truth does not hold, despite realizing that numbers around the 'half way' value cause problems. In fact, their assumption of the truth of the statement to 1 decimal place is correct. By (**24**) Claire is imputing a purpose to the problem, elaborated later (**27**), which encapsulates their interpretation and solution but is nowhere evident in the original question. 'Generalizable?' is thus taken as an instruction to generalize the problem statement having established its truth within their explicit bounds (**14**). 'Proof?' is interpreted as verifying this generalized statement, rather than their 'established truth'. One might want to criticize their understanding of the rigour of proof needed, but their ability here would be dependent on previous teaching and experience which may be lacking.

This whole area of the effect of pupils' own reformation of language and use of synonyms within problem solving, is an area worth exploring in greater depth than is possible with this one transcript. Two illustrations are given here of how mathematical symbolization forces the pupils to re-phrase their ideas. The problem revolves around numbers which are 'between' two other numbers, but to symbolize this relationship the pupils have to make the jump to equivalence with the phrase 'is smaller than' (**21**). A second incident occurs later, when needing more precision as to the relative proximity of the numbers, 'closer to' (in the question) becomes 'nearer than/to' (**22**), which then becomes an expression involving 'is less than'. Note that the same symbol, $<$, is used in both cases, although smaller and less are not grammatically synonymous in general English.

Practical implications

The analysis of this one discussion has illustrated the need to be aware of the pupils' interpretations of problems which they are solving before passing judgement on understanding and ability. It is necessary, too, to be conscious of the danger not only of pupils failing to understand the mathematics register as presented by the teacher, but also unknowingly using different interpretations of the 'obvious' language they use with their peers. If pupils are to spend more time in the classroom working in groups, then the 'rules' governing group work (accepted but unspoken by the pupils) must be brought out into the open and the teacher must be clear whether the aim is to have the pupils constantly verbalizing their mathematics (as these pupils were) or to encourage the use of the ideas of their peers, together with personal jottings, experimentation etc., in the solution of mathematical problems. Is cooperative, communal working or the solution of a mathematical problem the *main* goal?

Conclusions

This chapter has raised more questions than it can answer. Its aim was to illustrate, through detailed study of one transcript, the effect of group problem solving on the language the pupils used and, conversely, the effect their language had on the solution of the problem. The intention was not to present statistically supported generalizations, but through close qualitative analysis to offer paths of inquiry worthy of study if we are to untangle the connections between language use and mathematical understanding. It should be clear from foregoing comments that peer group discussion is not necessarily seen as the panacea to all errors and misunderstandings, but that if we are to move ahead in this field we must study with care the discourse between pupils in the natural settings where it occurs and not only that in which the teacher is involved. The final caveat must go to Duncker (1945), however: 'A protocol is relatively reliable only for what it positively contains, but not for that which it omits.'

Acknowledgement

I should like to express my gratitude to Rolph Schwarzenberger for his perspicacity and constant willingness to critique my writings.

14 Children talking in computer environments: New insights into the role of discussion in mathematics learning

Celia Hoyles, Rosamund Sutherland and Lulu Healy

In this chapter, we examine the ways in which pupils talk in computer environments. We begin with a summary of four aspects of the role of discussion in learning. With these in mind, we outline some of the issues that arise when we consider discussion within the context of collaborative interaction with computers in mathematical work. Then we present in some detail an analysis of the discourse processes whereby pairs of pupils come to make generalizations and to formalize them. We point out that different computer environments promote different modes of discussion and generalization. We consider some of the practical considerations of work on these topics, and conclude with a speculative, but we believe instructive, proposal about the ways in which interaction with computers may promote development.

The role of discussion in learning

From the literature, we can distinguish four interrelated aspects of the potential role of discussion in learning: *distancing*, *conflict*, *scaffolding* and *monitoring*. We find this classification useful in assisting us to identify significant aspects of pupil discussion in various computer environments. First, let us consider the notion of *distancing*. It is reasonable to argue that the opportunity to articulate one's thoughts helps to sort them out. Talking provokes a representation of one's thoughts – a process which inevitably raises them to a more conscious plane of awareness so that they can become the objects of reflection and modification. This is the cognitive component of discussion. This process of distancing from ill-defined intuitions is likely to be enhanced in a social situation where there is a need to communicate with others and therefore make ideas more explicit (for further elaboration, see Hoyles, 1985). Discussion of one's actions is a particular example of distancing. It can assist in uncovering the rules regulating the actions, since the use of a language involves 'a degree of semantic independence from the direct features of the environment' (Blundell 1988: 19; see also Olson *et al.*, 1987).

A second role for discussion relates to *conflict*. Some see discussion between peers as a mechanism to provoke cognitive conflict – a clash between different perspectives which can provoke a re-evaluation of initial perceptions (Perret-Clermont, 1980). Mugny *et al.* (1981) suggest that both social and cognitive conflict are necessary – but not sufficient – conditions for the efficacy (in terms of cognitive development of the individual) of a social interaction. In addition, in socio-cognitive developmental theory (Doise and Mugny, 1984), the concept of group cognition is introduced as an intermediate experience between a child's intuitive understanding and more analytic understanding. Given that this implies a move towards disembedded thought, it obviously relates to distancing.

The theories of Vygotsky (1962) support the demands for more pupil talk from a rather different standpoint and this brings us to the *scaffolding* role of discussion. For Vygotsky, individual conscious awareness is an internalized outcome of activity in a social context. He asserts that individual cognitive structures are actually formed through social interaction, and development constitutes a move from an inter-psychological to an intra-psychological plane. The zone of proximal development is described as the distance between what one can do alone and what one can do with assistance or in collaboration with 'more capable' others (see Vygotsky, 1978: 85–6). Within this theoretical framework, discussion has a scaffolding role in relation to specific tasks or understandings. The group shapes the behaviour of its members by throwing up lines of development and ways of reasoning which can be understood but not constructed by an individual alone.

Finally, let us turn to the *monitoring* role of discussion. It is reasonable to conjecture that talk facilitates metacognition, that is pupils' internal self-regulation and self-reflection upon the state of understanding of their learning processes. However, we suggest the monitoring role of discussion is also significant in the regulation and direction of the activity of the group – that is, in developing plans through the interchange of ideas, in monitoring and checking each other's actions, in keeping on track in relation to the goal of the activity and finally in explaining and convincing others in the group.

Finally, it must be said that despite considerable research effort related to group work and discussion, the absence of any epistemological focus renders most of the findings of limited value for our purposes. We propose some crucial questions as far as mathematics learning is concerned: Can discussion serve as a bridge from a focus on actions or specific cases to the criteria that govern them – criteria which reflect mathematical structures and relationships? Can discussion 'push' participants towards generalization and abstraction – processes at the heart of mathematics?

Discussion and learning in the context of computer activities

If the computer is introduced as a variable into the learning situation, we find that there is no significant body of research on group work with computers and little knowledge as to exactly the nature of the role of the computer in relation to peer interactions. It has been argued that computer-based activities 'invite' collaboration (Shiengold, 1987: 204), but collaboration to what end and why might this be the case? It may be, for example, that the computer provides a necessary focus for the discussion because of the public character of the screen. Or, is the crucial point the public nature of the action at the keyboard and the need to make a joint decision before this action can take place, as only one action can take place at a time? Or does the influence of the computer stem from the interactive nature of the environment and the feedback obtained? Much more work needs to be undertaken to address these questions as well as to identify the different patterns of interaction which facilitate certain types of learning and how these are influenced by different types of software and different types of group organization.* The point to stress here is that the computer is *not* a unitary object – its effects depend on both software and activity.

There is, therefore, a need for microanalysis of all the interactions in a computer context, i.e. inter-pupil and pupil–computer, in order to illuminate the interactional process. We need to know the outcome of the activity but also the processes by which this outcome has been achieved. We need to distinguish what the pupils are talking about – screen output, the underlying mathematical issues or the links between the two. One would hope that such an analysis might uncover significant incidents and critical decision points whereby pupils come to experience their mathematics in a different way and to see it in a new light.

In the *Logo Maths Project* (Hoyles and Sutherland, 1989), we started to address some of these questions through the analysis of the transcripts of pairs of pupils working with Logo over a period of 3 years. We found that, despite marked variation in patterns of interaction between pupil pairs, collaborative exchanges provided challenging ideas for projects and widened the range of projects chosen; kept a project going in the face of 'obstacles'; provoked appropriate changes in representation of the task; and, finally, facilitated the development of more flexible approaches to problem solving and programming. Discussion also had a motivational and affective pay-off which gave pupils a sense of ownership of the mathematics. More significantly for the purposes of this chapter, although initially the talk was action orientated, later in the research we identified individual conceptual development as a result of the three-way interaction between a pupil pair and the computer. In such cases, the graphical feedback of the computer provoked reconsideration of ideas and this, together with the symbolic representation provided by the Logo program and the pupil

*These questions are to be addressed in a research project 'Groupwork with Computers' (1989–92) funded by the Economic and Social Research Council and based at the Institute of Education, University of London, and the University of Sussex (Eraut and Hoyles, in prep.).

discussion, provided 'scaffolding' for pupils to move on from an earlier conception. Such episodes also tended to provoke more supported argument between pupils.

Finally, in any consideration of discussion between peers, it must be borne in mind that there is evidence of individual differences among children in their discussion styles (Durkin, 1986: 228). This is a general issue but possibly accentuated in a computer environment where there is an additional mode of interaction. Gender differences have also been widely reported (see, e.g. Hoyles, 1988). Boys not only tend to monopolize computer hardware but also find it difficult to share interactions at the keyboard and see arguments with peers as time-consuming and diversionary. Girls, on the other hand, tend to appreciate the opportunity available in computer-based group work for mutual help and the sharing of ideas.

Generalization and representation

We now turn to an ongoing research project (Sutherland *et al.*, 1989) concerned with an analysis of the way pupils working in pairs make generalizations and come to formalize them in mathematics. Our hypothesis is that peer discussion and the distinctive nature of an interactive computer environment will assist in these processes. The research is undertaken within a spreadsheet environment (EXCEL), a Logo programming environment (Turtle Graphics), and a paper-and-pencil environment. The aims of the research are to:

- Investigate the interrelationship between the negotiation of the generalization by the pupil pair and its formal representation.
- Investigate any effects on pupil response of the problem-solving tools made available by the different environments.

We have identified critical decision points in the generalization process and intend to analyse at these points whether or not decisions are jointly made, suggestions are justified by verbal explanation and disagreements occur.

One research task has been designed for each of the three environments. The common element of each task is the requirement to identify and formalize relationships embedded within specific cases defined by both visual images and numbers. It is intended that these relationships are not immediately obvious to the pupils (e.g. more complex than 'times by 2'). All the tasks have a well-defined goal but the process of problem solution is not specified. In particular, there is some element of choice as to the variables which can be used for defining the generalization.

In this chapter, we focus on pupil discussion while searching for mathematical generalizations and their formalization in the two computer environments. We are in the process of developing categories of pupil language designated as modes S, P, I, L, F, W and T (see Table 14.1) in order to describe the generalization process and to identify different patterns of pupil response between environ-

Table 14.1 Modes of response in the generalization process

Mode S	Making sense of the task, playing with numbers, brainstorming, questioning conjectures
Mode P	Seeing a pattern, and describing the pattern with reference to a specific case or cases
Mode I	Intermediate mode between P and L
Mode L	Expressing in natural language a way which describes the process by which the generalization can be obtained. Selecting a variable as a result of seeing a general pattern
Mode F	Expressing the processes of generalization in the particular formal language appropriate to the working environment
Mode W	Writing down the generalization on paper (this is the same as Mode F for the paper-and-pencil task)
Mode T	Trying out a specific case for an already defined formalization

ments. Our classification of pupil language by these modes is not only derived from the linguistic data in the transcripts but also from our interpretation (using video and observation) of the meanings and intentions underlying the pupil language. As such, the classification is inevitably subjective. None the less, we would argue that it provides a rich source of insight and hypotheses.

The following episodes serve to illustrate pupil discussion while searching for a generalization and its classification according to the modes. The transcripts are a pair of girls aged 12 years. The girls had already undertaken 7.5 hours of Logo work and had been introduced to writing and editing procedures and using and operating on variables. They also had experienced about 2.5 hours of spreadsheet preliminary activities involving entering data and formulae and replicating formulae containing relative references.

Spreadsheet examples: Penny and Nadine – polygon numbers

The task for Penny and Nadine was to generate the polygon numbers on a spreadsheet having been given a few starting numbers and a figurative description of these numbers as shown in Fig. 14.1. The pair had generated without difficulty the sequence of natural numbers to define the 'position' by replicating the rule 'add 1 to the cell before'. We first describe their construction of the triangle number sequence.

Penny and Nadine enter 1 in cell B2 and pause for discussion with the cursor on cell C2 (as illustrated in Fig. 14.2). Their initial strategy is to attempt to define a recurrence relation in which each term of the triangle number sequence is generated solely on the basis of the *term before* it (e.g. the 5th term is defined in

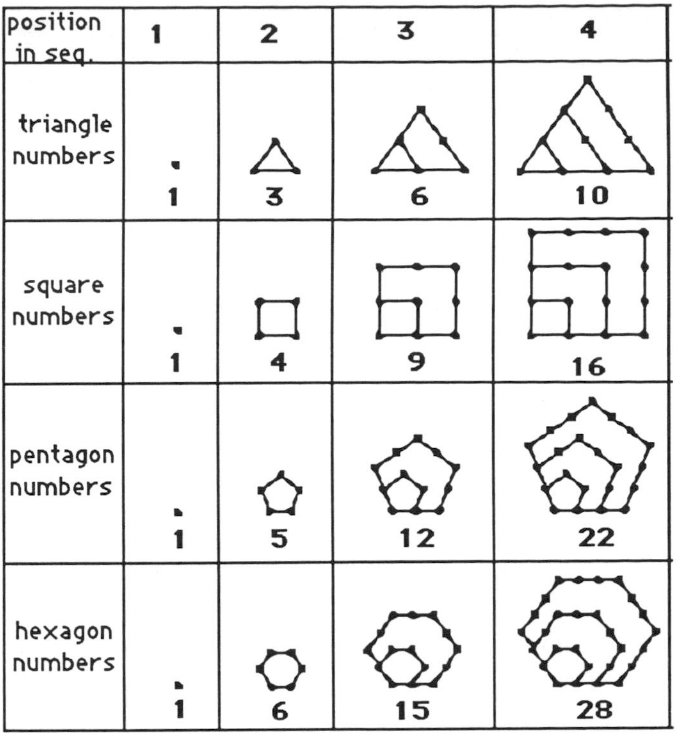

position in seq.	1	2	3	4
triangle numbers	1	3	6	10
square numbers	1	4	9	16
pentagon numbers	1	5	12	22
hexagon numbers	1	6	15	28

Fig. 14.1. Polygon numbers.

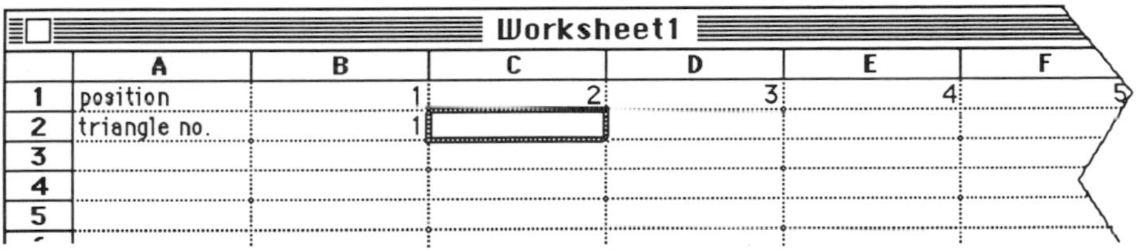

Worksheet1

	A	B	C	D	E	F
1	position	1	2	3	4	5
2	triangle no.	1				
3						
4						
5						

Fig. 14.2. Beginning the triangle number sequence.

terms of the 4th term). They are making sense of the problem (categorized by mode S) and in so doing come up with a number of hypotheses. After unsuccessfully looking for a rule using this strategy, Penny starts along a new path (still mode S) and tries out rules which derive from the *position* of the triangle number in the sequence. She then moves between these two strategies, looking at both the term before and the position in the sequence. Suddenly she stops, pauses, and announces her discovery of a pattern which synthesizes both approaches (a critical decision point). She describes this by reference to

Worksheet1

	A	B	C	D	E	F
1	position	1	2	3	4	5
2	triangle no.	1	=B2+C1			
3						
4						
5						

Fig. 14.3. The rule for triangle numbers.

particular cases, but uses actions (pointing to the cells on the screen) which suggest the way to generalize (classified as mode I):

> P: *1 2 3, hey look, look, I've found something look, here to get the next number you add 2* [meaning the move from cell B2 to cell C2], *here to get the next number you add 3* [meaning the move from cell C2 to Cell D2], *and there to get the next one you can add 4.* (mode I)

Through her simultaneous use of language and actions, Penny identifies how the sequence is generated. She then needs to formalize and articulates the formula as it would be generated on the spreadsheet:

> P: *Ah I've got, it's this one add this one, that one add that one equals that one, that one add that one equals that one.* (mode L)

Although her language does not sound like a generalization, while Penny is talking she is pointing out on the spreadsheet the *action* which when replicated, or generalized, will produce the sequence – that is, to obtain, for example, the value in C2 you add B2 (the cell before) to C1 (the cell above). There is then only one small step to formalization and Penny is able to enter her rule into the computer without difficulty or further comment as shown in Fig. 14.3.

Nadine, meanwhile, has not yet seen how the formalization produces the sequence of triangle numbers. She seeks clarification of Penny's generalization by reference to specific cases, but in a way that indicates that she understands the replication process.

> N: *Yeah but then it'll be doing that* [points at B2] *by that* [points at C1] *all the time* [carries on the pattern, pointing at C2 and D1, D2 and E1 E2 and F1]. *Is that what you want?* (mode T)

Nadine's query provokes Penny to explain her ideas by reference to specific cases and the worksheet. They then copy the rule across as illustrated in Fig. 14.4.

In the above example, Penny worked out the final formalization without apparent assistance from her partner. The articulation of her hypotheses and counter-examples may, however, have been provoked by Nadine's presence and this may have helped her to see the pattern through the process of distancing. Nadine's query near the end of the episode served a monitoring function for the pair and Penny's final explanation could be seen to provide scaffolding for Nadine.

Worksheet1

	A	B	C	D	E	F
1	position	1	2	3	4	5
2	triangle no.	1	3	6	10	
3						
4						

Worksheet1

	A	B	C	D	E	F
1	position	1	=B1+1	=C1+1	=D1+1	=E1+1
2	triangle no.	1	=B2+C1	=C2+D1	=D2+E1	=E2+F1
3						
4						
5						

Fig. 14.4. Generating the sequence of triangle numbers.

Worksheet1

	A	B	C	D	E	F
1	position	1	2	3	4	5
2	triangle no.	1	3	6	10	1
3	square no.	1	4	9	16	
4	pentagon no.	1				
5	hexagon no.	1				

Fig. 14.5. Beginning the pentagon numbers sequence.

Later on in the same session, when the pair was attempting to generate the pentagon numbers (having constructed the square numbers), collaboration between the pair took on a more significant role. The girls had realized that every number sequence began with 1 and had entered these. The spreadsheet they were working on is shown in Fig. 14.5.

The pair's initial strategy is again to look for recurrence relationships using previous numbers in the sequence of pentagon numbers (categorized by mode S). Penny then notices some new patterns *going down* the sheet. *This is a major shift in perception of the problem.* Her first description of the pattern uses mode P.

P: *Hey look, look 1 1 1 1 1, 2 3 4, 3 6 9, 4 10 16, 5 15 25, we can do it from that now, we can figure something out . . . so 2 3 4, that'll be 5, that'll be 5, then that'll be 6, yeh look, look, oh god it's so simple man, look at the sheet, just look 2 3 4 5 6 . . .* (mode P)

Nadine is trying to keep track of what is going on and Penny, in trying to explain to her, changes from describing the pattern with reference to specific numbers to a partial generalization (mode I):

P: *So it's add 1 to the number before it* [referring to column C], *then add 3 to the number before it* [referring to column D] *and that's add* . . .

N: *How do we get that into a formula?* (mode I)

Nadine's query articulates the aim of their activity and their problem. Neither girl can see a way of converting the various patterns going down the worksheet into a single rule which can be copied. They cannot make the transition to a general formulation to construct the pentagon numbers (mode L), nor can they see how to formalize their description on a spreadsheet (mode F). After a pause, Nadine points to the link between the pentagon numbers and triangle numbers with reference to specific cases.

N: *Umm, look it goes that* [indicates 4 in C3] *add 1 then that* [indicates 3 in C2-triangle number] *that* [the 3 in D1-position], *no, yeh but look the connection, that's 1, that's 3, that's 6* [points at triangle numbers]. (mode P)

Penny is able to make use of Nadine's observation and develop it to both find a pattern that works and express it as a generalization. Penny's language is not precise in the following extract – when she says 'number before', she is actually referring to the *triangle* number before. There is no misunderstanding between the pair but we insert [triangle] for clarity. Penny first rearticulates the idea:

P: *That's, that's easy, we just do it how we did before, so it's that number* [points at cell C1-position] *add 1* [points at cell B2-triangle number], *add the one below and across, add the one before, the position add the number before, the position . . .* (mode S)

At this point she realizes there is another obstacle to get over in order to make the generalization of their strategy, namely 'how far down you are'.

P: *Add, oh hang on, the position add the* [triangle] *number before, times it by however far down you are, times it by 4 cos you're 4 down* . . . (mode L)

She debugs this by testing on a specific case and finally suggests:

P: *So 3 times the* [triangle] *number before equals 3 add to the position . . . let's, shall we just try it.* (mode L)

Having now expressed the pattern in natural language in a way which matches the action on the spreadsheet by which the generalization can be obtained, the work has been done and the pair have no difficulty in entering the rule in spreadsheet syntax (Fig. 14.6) and copying the rule across.

In this episode, Nadine's first question served a monitoring function in keeping the pair on track and her later introduction of triangle numbers served to scaffold the activity for Penny. Penny's first (incorrect) generalization seemed to serve a distancing function in that after describing the pattern she paused ('oh hang on'), reflected and then debugged.

Worksheet1

	A	B	C	D	E	F
1	position	1	2	3	4	5
2	triangle no.	1	3	6	10	1
3	square no.	1	4	9	16	
4	pentagon no.	1	=(3*B2)+C1			
5	hexagon no.	1				

Fig. 14.6. The rule for pentagon numbers.

Logo example: Penny and Nadine – P and N design

Let us now turn to a Logo example. Penny and Nadine had already defined a procedure PP, which produced a variable size P (Fig. 14.7a) and had chosen their input SIZE for length DC (a critical decision point). They were then working on a design of their own to incorporate an N (for Nadine) into the figure (see Fig. 14.7b). They had called up the procedure PP with an input of 45 and found that the length AE was 50 by estimation – trying a number and then improving it on the basis of visual feedback.

In order to write a new procedure PANDN, the pair had decided to copy out the commands they had used in PP (where the turtle ends at A) and then add on new ones. The first two distances AD and DE did not cause any problems because they had already worked out how to operate on SIZE to obtain them, i.e. 30 is (SIZE/3)*2 and 45 is SIZE. The episode begins at the point when they are trying to deal with the length AE of 50.

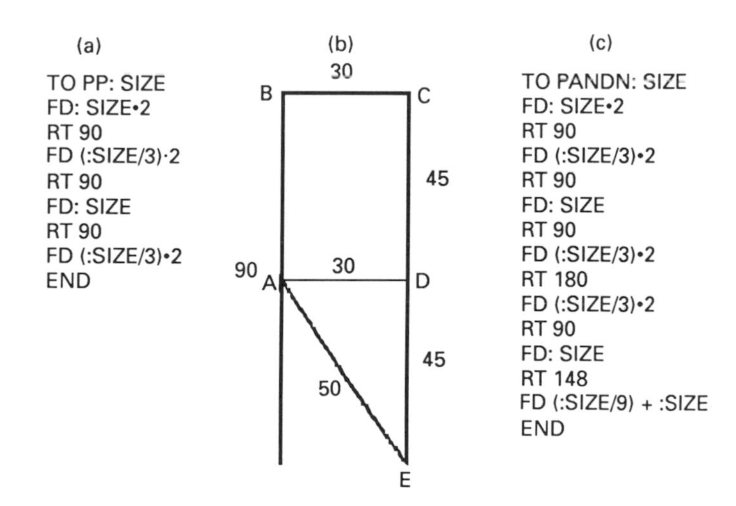
(a)
TO PP: SIZE
FD: SIZE•2
RT 90
FD (:SIZE/3)·2
RT 90
FD: SIZE
RT 90
FD (:SIZE/3)•2
END
(b)
30
B
C
45
90
A
30
D
45
50
E
(c)
TO PANDN: SIZE
FD: SIZE•2
RT 90
FD (:SIZE/3)•2
RT 90
FD: SIZE
RT 90
FD (:SIZE/3)•2
RT 180
FD (:SIZE/3)•2
RT 90
FD: SIZE
RT 148
FD (:SIZE/9) + :SIZE
END

Fig. 14.7. The P and N design.

Penny suggests an additive strategy in the formal language appropriate to the Logo environment and Nadine goes along with it:

P: *OK, and fd 50 equals maybe size add 5 . . .* (mode F)

However, immediately the assertion is made, Panny rejects it. She *sees* that in general, for different values of SIZE, adding could not work:

P: *But no that wouldn't work for everything would it?*

She then tries to come up with some hypotheses concerning the relationship between 45 and 50.

P: *OK 45, 45 is the size for this yeh, so 45 does what to get 50?* (mode S)

Nadine then makes a significant comment which introduces into the discussion the idea of 5 as a 'unit':

N: *You could use 5's into that . . . use 5's . . .* (mode S)

At this point, Penny ignores Nadine's contribution, but later she does take it up:

P: *What's 5 a percentage of, I mean how many times does 5 go into 45 . . . 9 times OK, so . . .* (mode P)

Penny now sees a relationship between 5 and 45 which she partially generalizes:

P: *Say size is 45, then say 5 goes into 45 nine times, so it's, size divided by 9 . . .* (mode I)

Nadine obviously still does not understand why it is necessary to multiply rather than add.

N: *Add 5 no?*
P: *No size divided by 9.*
N: *Add 1 is ten, no.*

Penny ignores this and states the generalization again using the formal language of Logo:

P: *Divided by 9 . . . add size, size divided by 9, add size.* (mode F)

The pair then finish typing into the editor the procedure PANDN (Fig. 14.7c), test it out and find that it works.

In this episode, it is noteworthy that in the construction of the formalization the pair moved straight from spotting a method that worked for one specific case to expressing it formally using Logo syntax. It is reasonable to conjecture that the mode F response served as scaffolding towards generalization. Penny's first suggestion of the addition strategy to obtain length AE seemed also to serve for her a distancing role leading immediately to cognitive conflict and a rejection of her conjecture. Nadine's contribution is the introduction of a method which could be used to obtain a multiplicative relationship between the variable and the required length. It served in a scaffolding role towards solution.

Synthesis

Generalization in the two computer environments

In this Logo environment, pupils tend to generalize by using a substitution strategy, that is by thinking of a generalization in terms of a specific number and then substituting the variable for this number. Here, for example, the pupils think of SIZE in terms of 45 and work out the operations on 45 in order to come up with general relationships. Once these are defined, the pupils can experience the variability of SIZE while interacting on the computer. Naming the variable in the Logo formalization is important (see Sutherland, in press) but, additionally, when in pairs, pupils from the outset communicate in Logo-specific language, mode F, simultaneously with the other modes of response. Using mode F language gives a means of articulating general ideas and therefore provides pupils with scaffolding, enabling them to view the specific case as a generic example. The internal relationships are worked out with respect to this specific case, but with an underlying awareness that the relationships have to hold for all numbers. Pupils can play with their procedures, switching between visual and symbolic representations in the process of groping towards a fuller understanding of the formalization. Their discussion is part of this process.

In the spreadsheet environment, the software itself does not provide a means for talking about possible generalizations since the syntax refers to specific cells. It is the relationship between the cells, expressed by the replication process, which carries the generality. Thus negotiation with the software is through action rather than spreadsheet-specific talk. Inter-pupil discussion assists in seeing a pattern by distancing pupils from their actions. However, the natural language used is modes S, I and L, rather than the formal spreadsheet language, mode F. The way the generalization is constructed and expressed in natural language is certainly influenced by the computer environment. The language matches the actions on the spreadsheet and is only meaningful with reference to the spreadsheet environment. This means that the natural language formulation essentially solves the problem. The words used drive the actions to be undertaken on the spreadsheet and these actions reflect the necessary mathematical relationships. The spreadsheet itself then takes care of the details of the formalization. Mode F language therefore does not occur, or if it does it is at the end of the solution process rather than at the beginning – in contrast to the Logo environment.

Thus we suggest that with Logo the natural language of the pupils and the software tools develop together in a dialectical way and serve simultaneously as scaffolding towards generalization (see also Hoyles and Noss, 1987, for discussion of the scaffolding role of Logo). If provisional generalizations are made, the visual feedback can convince or give assistance with debugging. With spreadsheets, the influences of inter-pupil and pupil–computer interaction operate at different points in the generalization process. Inter-pupil discussion assists in the spotting of the pattern and its specification in terms of spreadsheet actions. Pupils do not seem to use the numerical output available on a

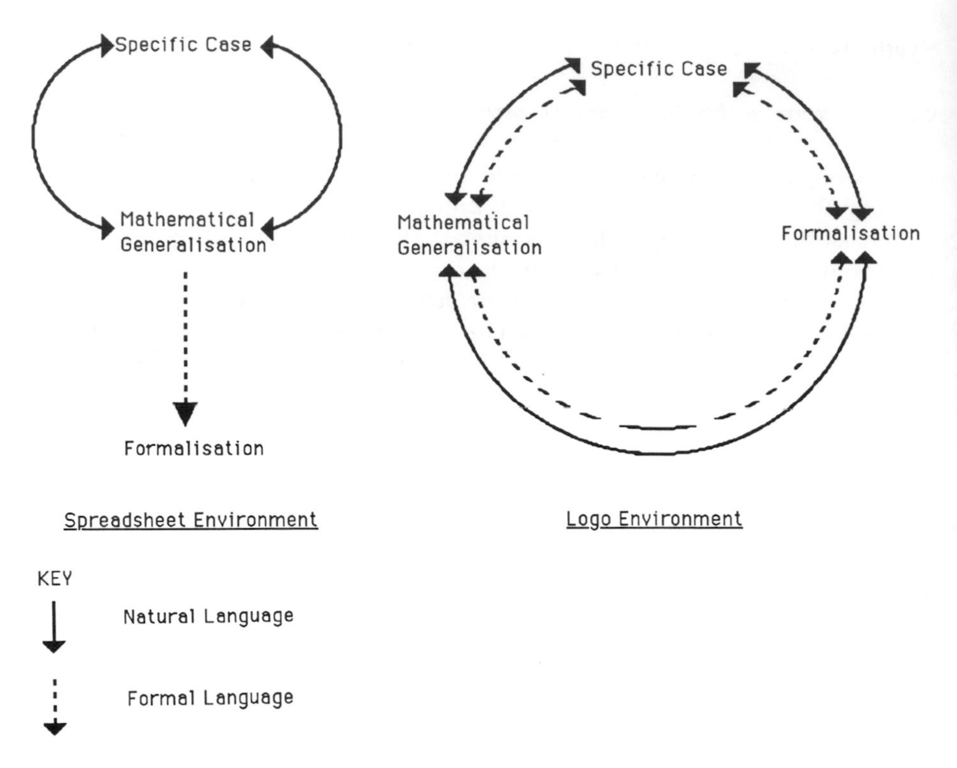

Fig. 14.8. Natural and formal language use in the generalization process in spreadsheet and Logo environments.

spreadsheet to assist them in this process but, having completed it, the software tools facilitate formalization. Figure 14.8 illustrates the two contrasting situations.

Practical implications

Computer contexts provoke active discussion
We have seen that both computer contexts provoked the pupils to work in an active way and to talk extensively and thoughtfully about the mathematics problems involved. Significantly, the talk was almost exclusively on task except when the work was very difficult, very easy or repetitive. This is in contrast to our findings in paper-and-pencil environments.

Different computer environments promote different types of discussion
In both computer contexts, the language of the pupils served as a bridge between a pattern seen with respect to a specific case and its generalization in the appropriate formal language of the working computer environment. However,

there are striking differences in the articulation of the generalization process. These differences closely relate to the tools available in the different software environments. Teachers therefore need to be aware that the choice of software environment and the choice of task are important decisions affecting the style of interaction pupils are likely to adopt.

Concluding remarks

Our work suggests that pupil language by reason of distancing, monitoring and scaffolding can play a critical role in overcoming obstacles during the solution process. When activities are comparatively straightforward, it seems that collaboration has social rather than cognitive significance. We also suggest that the software tools themselves can serve a scaffolding role in relation to the solution of the mathematical tasks. Finally, it is noticeable that for this pair, Penny and Nadine, there is very little disagreement or social conflict. This may simply reflect the pair's style of working together and this itself may possibly be an example of a broader gender preference. We accept that other pairs may have rather different profiles of interactions. Nevertheless, we conjecture that when pupils work together in a computer environment, their discussions influence the interaction with the computer, but it is the interaction with the computer which provokes any cognitive conflict necessary for individual conceptual development. Where programming is involved, this conflict will arise at the point where there is a need to write formal code, a point which varies across software environments. It is at this instant that inter-pupil and intra-pupil incompatibilities and inconsistencies as to perception of the problem and structure of the solution *have* to be negotiated, made explicit and then ultimately made to reflect the mathematical constraints of the task.

Further reading

Hoyles, C. and Sutherland, R. (1989). *Logo mathematics in the classroom*. London: Routledge.

Recherches en Didactique de Mathematique, **8** (1.2) (1987). Special Logo edition.

Section 6

Language, mathematics and disability

15 Mathematics and the deaf child

Jeffrey Barham and Alan Bishop

Introduction

Most teachers of deaf * children, at both primary and secondary level, would agree that mathematics is one subject with which their pupils find special difficulty. Yet it is not immediately clear why this might be so. Indeed, because it is a subject which contains much visual figural material, and which relies heavily on the use of symbolic representation, it might be considered a subject where deaf children should find their disability no handicap. Such an attitude might be typified by a remark frequently made to the authors: 'You don't have to hear to add up, do you?'

Nevertheless, mathematics does seem to be a subject which many teachers of deaf children will insist contains more than its share of problems for those children. Any visitor to a unit for hearing-impaired youngsters will see them working at a lower level than their hearing peers. This impression will be enhanced by a glance at the materials and books being used, which will usually be those specifically designed for much younger children.

There is a great deal of research evidence to suggest that the far from encouraging impression we began with is a valid one. Hine (1970) tested the attainments of over 100 deaf children, aged between 7 and 16 years, by giving them a standardized test (Schonell's Essential Mechanical Problem Arithmetic Test, 1960). Hine (1970: 134) summarizes the results by stating that:

> It can be concluded, therefore, that at age 10 these partially-hearing children have an average attainment of about $8\frac{1}{2}$ years, while by the age of 15 this has reached about $10\frac{1}{2}$ years in mechanical arithmetic and about 11 years in problem arithmetic.

*We shall use the word 'deaf' generally to refer to any child suffering a hearing-impairment, while being fully aware of the range of impairment which does exist among children.

Hine (1970: 133) further notes that between the ages of 9 and 11, the children tested appeared to make no appreciable progress at all.

Austin (1975: 360) explored the development of deaf students aged 10 to 20 years through 'the measurement of their knowledge of selected concepts related to independent living'. Two of the eight concepts have a bearing on mathematical attainment – money and measuring. The results obtained supported Austin's (1975: 361) hypothesis that 'the percentage of concept knowledge demonstrated by the deaf population at age 20 will be exceeded by hearing students at age 14'.

Newman (1982: 37), discussing the work of the prestigious Gallaudet College in Washington, D.C., has to admit:

> Even those deaf students who have normal to above average I.Q.s are not functioning on a high level in a subject that should be their greatest strength – mathematics.

Wood *et al.* (1986), using the Vernon-Miller Graded Arithmetic-Mathematics Test with a total of 1005 children of whom 540 were hearing-impaired, found the following data:

Type of school	*Maths age*	*Type of school*	*Maths age*
Special schools	12.1	Mainstream	14.0
Partially hearing units	12.8	Hearing children	15.5

It does seem, therefore, that deaf children consistently appear to maintain levels of achievement in mathematics which one might expect from either younger or less-able students. This must certainly affect not only their self-esteem but also the expectations of their parents and of their worth in a generally unsympathetic and competitive society. Tomlinson-Keasey and Kelly (1974: 693) comment upon the 'persistent and frustrating feeling . . . [by countless teachers of the deaf] . . . that their students' full potential will never be realised'.

How then can we understand this under-achievement, and what might a mathematical education offer to the deaf child?

Language difficulties in mathematics for deaf children

When asked about the problems their pupils are having with mathematics, teachers of deaf children seem to have an intuitive feeling that language is at the heart of their difficulties. Certainly, one has only to hear deaf children speak, or examine their written work, to recognize some of the effects of hearing impairment, and despite what has been said about the use of figural and symbolic material in mathematics, it is nevertheless a school subject which involves a considerable use of language, often in ways not helpful to any language-impaired child.

There are a number of ways in which the language of mathematics can cause problems for deaf children who lack the degree of vocabulary enrichment of their hearing friends. First, even at a very early stage, the child is introduced to new and potentially confusing words, e.g. 'multiply' and 'rectangle'. Often there is confusion between words which might sound similar, especially to a child who is employing lip-reading, e.g. 'ten' and 'tenths', 'sixty' and 'sixteen'. Words may also be used in mathematics in a different way to which they are used outside the classroom. A teacher of deaf children will endeavour to teach the concepts of 'up' and 'over' to a child who will then meet the words in such phrases as 'add up' and 'left over'. In this respect, deaf children are faced with a particularly marked version of a general ambiguity inherent in spatial terms in mathematics education, as discussed by Durkin and Shire (this volume). Nickerson *et al.* (1986: 17) list a number of phrases commonly used in examinations which might cause confusion to hearing-impaired students. For instance, 'bring to rest', meaning simply to 'stop', might be interpreted as 'go and have a sleep', whereas 'overall', the measurement of everything, might be seen as an article of clothing.

Another area of confusion for deaf children is the fact that in mathematics there may be more than one way of expressing the same idea. Preston (1978) examined a number of books commonly used in primary mathematics classes and found 18 ways in which a simple addition sum might be 'read aloud'. Teachers of deaf children often find great difficulty in getting a child to understand *one* way and are therefore haunted by the thought of a further 17 ways looking over their shoulders.

Deaf children, like all language-impaired youngsters, have particular difficulty with logical connectives, the (usually) small words which form the skeleton of much of our thinking and talking. Words like 'if' and 'because' are rarely taught specifically to hearing children. Although these terms do present semantic and syntactic difficulties even for normal language learning children (Donaldson, 1986), they have opportunities to learn them in everyday interactions. Deaf children have little access to this 'informal' learning and have to be taught the meaning of these words, and helped to understand the concepts they represent, slowly, deliberately and patiently. Silverman-Dresner and Guilfoyle (1972) tested several hundred hearing-impaired children in various American schools to assess the extent of their language development. They found that most children did not understand such words as 'if', 'because', 'most' and 'some' until they were well into their 'teens. Yet these words form the basis of much mathematical activity. A similar pattern was seen for many of the words specific to mathematics, e.g. 'add', 'amount', 'count', 'quarter' and 'subtract'.

Language and concept formation

Yet in mathematics, as in many other areas of human thought, language does more than simply provide a vocabulary. It is indeed basic to the formation of many concepts and mental processes.

Furth (1966) reports on his research which indicates that deaf children have more difficulty with the concept of opposition (things being different) than with the concept of similarity. It might well be that consideration of the degree of language used internally might provide a clue why this is so. Once one has recognized that two objects are the same, little more remains to be said, but in comparing different objects more language-based concepts need to be drawn upon.

Stone (1980) discusses the difficulty deaf children have with sequencing, i.e. putting things in order. Teachers of young deaf pupils have frequently told the authors about children who 'seem to know their numbers, but have difficulty putting them in order'. Again, a consideration of language might indicate why this is so, for once the number of objects has been appreciated, say the 'fiveness' of five, little more can be said. But to relate 5 to 4 and 6 requires the use of concepts such as 'greater than' or 'less than'. The authors found that the concept of sequencing caused problems not only to *young* deaf children. Teachers at the Mary Hare Grammar School (near Newbury, Berks, UK) told them of pupils working for 'O' (at 16+) and 'A' (at 18+) levels who, having done an experiment in science, had difficulty writing down the order in which they did things; biology students were not always able to immediately recall the stages in the development of an organism; mathematics students frequently avoided proof questions at examination times because they were unable to follow the steps of a theorem through to the conclusion.

Of course, an understanding of such language-based ideas as similarity/opposition, conservation and sequencing is vital to all mathematical activity from an infant school upwards, as other contributors to this volume demonstrate. However, Suppes (1974: 168) claims that:

> Cognitive performance of deaf children is as good as that of normal-hearing children, when the cognitive task does not directly involve . . . verbal skills.

But in most classroom mathematical activity, verbal skills *are* involved. The children must understand the teacher's question 'What happens if. . .', and they must discuss with their partners 'What do we do now?'

An appreciation of the difficulties facing the young deaf pupil in the mathematics classroom is further enhanced when one examines various language structures which might be used in both oral and written instruction. Rudner (1978: 33) lists a number of these used in a series of standard mathematical achievement tests which he gave to hearing-impaired students and which caused special difficulties:

- conditionals (*if*, *when*);
- comparatives (*greater than*, *the most*);
- negatives (*not*, *without*);
- inferentials (*should*, *could*, *because*, *since*);
- low information pronouns (*it*, *something*); and
- lengthy passages.

Characteristics of deaf children which might affect their mathematical thinking

Finally, deaf children may well exhibit difficulties which seem to be unrelated to their language impairment. Take, for example, the research of Dietz and Williams (Glennon, 1981), which indicated various traits among their pupils. They report, for instance (Glennon, 1981: 326), that certain emotional and behavioural problems are between three and six times more common with deaf children than with hearing children. Among such problems are:

- hyperactive lack of control: impulsive, unreflective and uninhibited behaviour;
- anxious inhibition;
- preoccupation and obsessive concentration; and
- aggressiveness, anxiety and hostile isolation.

Such findings underline what has become a truism among teachers of deaf children that 'the inability to hear is one of the least of their problems'.

Dietz and Williams further refer (Glennon, 1981: 336) to other problems that they have regularly met with hearing-impaired children in their classrooms:

- a lack of persistence in working through difficult problems and attempting to discover new relationships.
- an inability to apply what has been learned to life situations.
- an inability to transfer old learning to new.
- poor study habits.
- an inability to remember what has been learned.
- a tendency to have the learning of a new skill interfere with what has previously been learned.

We can now perhaps understand the increased difficulties experienced by deaf children in a normal mathematics classroom, particularly if we consider the work of the Russian psychologist Krutetskii, who sought (1976: chs 12–14) to identify certain factors that help make up an overall ability in mathematics. Among these, he includes:

1 The ability to generalize mathematical objects, relationships and operations. This requires knowing what to apply and where to apply it.
2 The ability to curtail the processes of mathematical reasoning and the corresponding system of operations. This means the ability to miss out steps in a mathematical argument.
3 Flexibility of mental processes. One must be able to switch freely and easily from one mental process to another and also be conscious of the diversity of possible approaches to solving the problem.
4 Striving for clarity, simplicity and economy (elegance) in a solution.
5 Reversibility of mental processes in mathematical reasoning – the ability to switch freely and rapidly from a direct to a reverse train of thought.

Are we saying then that mathematics is inaccessible to deaf children? Most definitely *not*. Neither are we suggesting that a mathematical education has nothing to offer deaf children. Indeed, we would take the view that good mathematics teaching has a great deal to contribute to a deaf child's education.

So what can be said about this 'good mathematics teaching'? We will restrict our focus to two aspects – the curriculum content and the use of microcomputers – to see what generalizable ideas can be obtained, of value to teachers in all kinds of schools who may be teaching hearing-impaired children.

Practical implications

The value of mathematics in the overall development of the deaf child

Of many points concerning the mathematics curriculum, the following have been selected as being the most significant:

1 When considering the mathematics curriculum for deaf children, there is a danger of conceiving it simply in terms of 'what *they* need to know'. It must be recognized, of course, that hearing-impaired youngsters, no less than others, require certain skills if they are to be able to live fulfilled lives in the harsh world outside the classroom, but any educational provision which risks differentiating between the disabled and able-bodied must be scrupulously avoided.

2 As well as what might be termed 'social' or relevant mathematics, the deaf child needs access to the fun and puzzle aspects of mathematics. One of the authors was recently shown, by a very experienced teacher of young deaf children, a scheme of work of which she boasted that everything in it would be 'of value to them when they leave school'. Certainly, that boast was not an idle one, but when she was asked what 'fun activities' the deaf children took part in during their mathematics lessons, the teacher replied that *she* had never found mathematics fun. The inference seemed to be that she saw no reason why her pupils should either!

3 Indeed, it may well be that the least overtly utilitarian of the activities enjoyed by most children in the mathematics classroom are the very ones which deaf children should be allowed to experience. A class of deaf children of our acquaintance placed a cane in the grass and every quarter of an hour, a wooden label with the time written on it was put along the sun's shadow. The next day, the children were able to tell the time, using their sundial. This activity, it might be claimed, had no possible value – there was a clock on the classroom wall and most of the children had watches. But the language that was generated, the excitement created, the pride that was felt as their hearing friends were shown what they had been doing, justified what might have been considered a 'useless' piece of work.

4 As Ling (1978: 47) reminds us:

> The vividness of the experience linked with the appropriate use of words spoken directly to the child are likely to be more effective than repetitive drill in a lesson that has been artificially devised to teach a new concept.

5 Spatial, figural and geometric activities certainly should not be omitted from the curriculum, in the assumption that only arithmetical work is of value. Indeed, the relationship between spatial and mathematical abilities is strong (see Bishop, 1980, for a summary of the relevant research) with developments in each contributing to the other. The mathematics of space is in any case arguably more important than that of number to any child.

6 Not only is mathematics of potential value in enriching the language of young deaf children, it is also capable of helping nurture skills which are important to the social and emotional development of such children. First, most teachers and parents will appreciate that hearing-impaired children often have difficulty, when faced with an unfamiliar situation, with shopping, standing back and considering the problem carefully. Mathematics, even at the infant class level, encourages reflection, looking at a situation and planning what we are going to do. The actions we take frequently have to be performed in a certain order for success to be achieved.

   Secondly, hearing-impaired youngsters often appear to have difficulty in understanding and appreciating cause and effect – seeing something happening as a result of something else. Mathematics, properly presented, provides ample opportunity for this to happen. It must be more than just the development of calculating skills or the learning of algorithms. Investigative activities, by definition, require children to see why things happen and to anticipate the results of what is done.

   Finally, there is a tendency for parents and teachers to 'spoil' and overprotect their deaf children with the result that the children never have to make any choices for themselves, and quickly develop what might be termed 'acquired helplessness'. However, mathematics has a great deal to do with making choices, whether it be to decide if an addition or subtraction sum is needed or what substitution to make to solve a differential equation. The mathematics curriculum can offer plenty of opportunities for 'decision-making' activities.

The microcomputer as an aid in teaching mathematics to deaf children

The authors have recently been engaged in a research project concerning this topic (see Barham, 1988) and it became clear through this research just how powerful an aid it can be to the teacher. In particular, the following points were significant:

1 Controlling the attention of deaf learners can be extremely difficult, but good microcomputer activities with a rich, visual, screen presentation can be very engaging. In our own work, it was striking to see how controlling the screen was – as one knows of course with hearing children also (see Hoyles *et al.*, this volume). Pupils did not have to look around the classroom to discover what was happening or to learn.

2 Successful programs seem to be those with in-built motivation, via control by

the children of what is happening on the screen. So rarely do they feel in control of their own activities, that they responded very positively to this feature. The researchers were able to witness the effectiveness of the computer as a means of promoting interaction, as they watched Debbie, new to school and initially bewildered by everything around her, change from a withdrawn little girl, whose only reaction was a wan smile, to one who became aware of her ability to succeed in a given task.

3 The highly visual nature of the activities meant that the children's language handicap was not so significant. Indeed, the task we set ourselves included creating activities using visual and figural representations of essentially linguistic and logical ideas, such as 'If . . . then' and 'Why . . . because'. Programs which fill the screen with words obviously do not have the same benefits.

4 The repetition possibilities of a microcomputer can be very powerful, enabling sequences, processes and events to be repeated in precisely the same form for however many times it is required. This enables the child to attend to the significant movements, and to learn to 'see' the required patterns without disturbance.

5 Our programs included different levels of difficulty, enabling the teacher to pitch the task initially at an appropriate level. It is crucially important for deaf children, who are often failed by our educational efforts, to experience success and for that success to be noticed, rewarded and built upon.

6 Our programs were created with the express purpose of allowing the teacher to control them, so that they could be stopped and discussion allowed to take place. Rather than merely 'plugging' the children into the captivating computer and leaving them alone, we felt it was important to use the microcomputer as the main activity-generator. This then freed the teacher to concentrate upon the language development and experience-reflecting activities which are so critical in the deaf child's education.

7 We used a variety of input devices with the microcomputer – keyboard, light pen, concept board – and it was striking to see how important movement was, particularly for the younger children. The light pen was a peculiarly significant device in that regard, and it appears to have benefits not always exploited by programmers. Also the children moved heads, arms, fingers and legs in their mental efforts to tackle the activities facing them. It is clearly important that such kinaesthetic 'modelling' be allowed and even encouraged.

Certainly, the advent of the microcomputer has provided ample opportunity for deaf children to be placed in situations where they do have to make choices, and to live with the choices they have made. They can now become masters of the situation, where they are in control and where success or failure depends upon what they choose to do.

Conclusions

We have therefore seen how mathematics, can be a valuable source of enrichment and stimulation, rather than passively suffering as a result of the deaf child's language impairment. It should be so much more than simply 'doing sums', no matter how vital such activity might be for the social welfare of the child. The French novelist, Victor Hugo asked: 'What matters deafness of the ear, when the mind hears? The one true deafness the incurable deafness, is that of the mind.' It cannot be claimed that mathematics will help deaf children to hear, but it will encourage them to think. By doing so, they may well learn to overcome the 'true deafness'.

16 Mathematics and dyslexia

Michael Thomson

Dyslexia literally means 'difficulty with words', and describes a severe difficulty with reading, writing and spelling in otherwise normal children. The term specific learning difficulty is often used synonymously, but essentially dyslexia refers to a language difficulty which primarily affects the written form.

As most contributions to this book demonstrate, learning mathematics is a highly verbal task, and one which from the outset calls upon the ability to cope with written language (see, especially, Sinclair, this volume). What happens, then, when the dyslexic child faces mathematical instruction? In this chapter, I will provide first a general description of the problems of the dyslexic child, and mention some key points concerning the aetiology of dyslexia that may bear on mathematical work. Then, I will present a number of illustrations of the kinds of difficulties that I and others have observed in the mathematical performance of dyslexic pupils. I will summarize the (scant) research to date on mathematics and dyslexia. Finally, I will provide an outline of several practical implications for teaching maths to dyslexics, based in part upon research findings and in part upon our own experiences with pupils at East Court School, a school for dyslexic boys and girls.

The nature of 'dyslexia'

General description

The following is a generalized description of the dyslexic child, as given by Thomson (1984, 1990). Typically, the dyslexic child can come from any socio-economic group, with a range of family attitudes towards school and learning. Developmental milestones will have been within normal limits, although language may well have been delayed. In particular, there may have been slight phonological or articulatory difficulties persisting beyond the 'norm'. The child may manifest these speech difficulties during the childhood years, particularly the confusion of (ø/f) and (r/w), and in some cases there may be some difficulty in expressing ideas fluently in spoken language. However, the child's use of language in relation to aspects of vocabulary, verbal reasoning, understanding of events, and other intellectual skills associated with higher linguistic functioning is within normal limits, or may well be exceptionally high. (However, it is a mistake to assume that all dyslexic children are highly intelligent.)

This point of discrepancy between the child's perfectly adequate and sometimes above-average oral ability in understanding, and his or her poorly developing written language skills is one of the first signs to parents and teachers that something is amiss. A child may be bright, intelligent, able to answer questions and (as far as parents and teachers are concerned) appear to have no apparent difficulty in coping with early aspects of school work. Over the first years, the child may be a bit slow in acquiring the alphabet and aspects of sight vocabulary, but parents may have been told not to worry as the child will develop in due course. By about the age of 7, the child has barely made a start at reading and spelling, and has become demotivated. As the child gets older, if he or she is not given appropriate help with reading, writing and spelling, these will lag further and further behind chronological age.

There may well be other reading and spelling problems in the child's family, either other diagnosed cases of dyslexia, or parents who have a spelling difficulty or who were late readers themselves. The child will have a reading problem. This can be characterized by slow and non-fluent reading, regressive eye movements, losing their place, reading from the middle of words or non-recognition of letters. Reading errors will include the confusion of letter order (e.g. bread for beard) and occasional reversals of letter or whole words (b/d, on/no), although these errors are not as frequent among dyslexics as commonly believed. Other errors include the telescoping of sounds and an inability to read phonologically: the child may be unable to sound out words correctly, to blend them, or to break them down into whole units.

As well as the above, the following are associated with dyslexia:

- A puzzling gap between written language skills and intelligence.
- Greater difficulty in spelling (production) than reading (reception).
- Difficulty in expressing ideas in written form.
- Use of bizarre spelling: some examples I have encountered include *raul* (for urchins), *kss* (snake), *fuda* (thunder), *iriteap* (terrific).
- Confusion of left/right direction.
- Sequencing difficulties such as saying months of the year in the wrong order; direction scan in reading; sequential memory.
- Poor short-term memory skills (repeating digits following complex instructions).
- Problems in acquiring arithmetical tables.
- Problems repeating polysyllabic words (sas'tis'tic'cal for statistical; per'rim'min'ery for preliminary).
- Difficulties in expressing ideas in written form.

Aetiology

The causes of dyslexia have basically been researched at two levels: first, the neurological level, in terms of brain function and, secondly, the cognitive level (the memorial, perceptual and other skills manifested by dyslexic children). Briefly, there is some evidence to indicate that there may be differences in the way the left hemisphere is organized in dealing with written language in the dyslexic. This could be important as far as mathematics is concerned, because, as we shall see later, the aspects of mathematics dyslexics find particularly difficult are concerned with verbal labelling or aspects of verbal short-term memory. This has been linked with aspects of left cerebral hemisphere processing. However, for practical purposes of everyday teaching, we need to look at the other level of explanation, that is in terms of the cognitive domain.

There is a good deal of research indicating that dyslexics have weak short-term memories, i.e. they have difficulty retaining remembered items over time. This applies particularly to material which is presented auditorily, or requires verbal labelling. In addition, dyslexics have problems with 'phonological coding'. This is the skill that is required to translate a visual symbol, such as a letter, into its name or sound. Thus difficulties in sound/symbol or grapheme/phoneme correspondence, and the associated development in learning the alphabet, can be seriously delayed owing to this problem of translating the visual symbol into its sound code. Phonemic awareness and syllabification are other areas of suggested weakness. Phonemic awareness refers to the ability of individuals to break down words into their constituent parts, such as 'dog' into 'd'-'o'-'g'. Similarly, the discrimination of sounds is particularly marked where two, three or four syllables are required, whether in reading or spelling (for reviews, see Snowling, 1987; Thomson, 1990).

Dyslexia and mathematics

The above difficulties, though obvious in the language domain, seem unrelated to mathematics. However, it has been established that some aspects of maths do create problems for dyslexics. We shall first look at some classroom observations and then examine what little research there is on the subject.

Classroom observations

Almost all descriptions of dyslexia include some reference to mathematical problems. Unfortunately, there are not always clear distinctions made within 'mathematics'. Mathematics is an umbrella term under which various sub-skills may be assumed, i.e. arithmetic, algebra and geometry. Other sub-classifications might include the four basic operations of arithmetic (i.e. addition, subtraction, division, multiplication), or measurements, extensions, fractions, and so on.

Thomson (1990) describes dyslexic children as having arithmetical problems with tables, basic operations and place value. Miles (1983) notes that 90% of dyslexics have particular difficulties with tables, as compared with 50% of controls. Joffe (1981, 1983) found that 60% of dyslexics were 18 months behind their chronological age in mathematics. However, she also noted that 11% of the dyslexics in her study actually excelled at mathematics. In our own studies at East Court School, we found that, initially, dyslexics were particularly poor at addition, subtraction, multiplication and division; however, they were often as good or better than non-dyslexics at measurements, the use of money and applying mathematical concepts to problems.

It should be stressed, however, that as with written language, there is little correlation between intelligence and mathematical skills as far as dyslexics are concerned. For example, Joffe (1981) found that there was no correlation between maths skills and IQ in her dyslexic group, but that there was a significant correlation in the control group. This again highlights the discrepancy between expected attainment (given chronological age and intelligence) and actual performance in dyslexics.

As far as addition and subtraction are concerned, many of the errors produced by dyslexics appear to centre around 'carrying' operations, which might be explained in terms of their weak short-term memories. Often, addends are misremembered and, occasionally, decimal place value is confused. Miles (1983) provides some interesting case histories of how children try to work out sums. For example, a boy who was asked to subtract 7 from 19, replied that the answer was 13: 'I took 9 away from 7 leaving 3 [*sic*] and added 10 makes 13.' This suggests that he had weak number sense, as $9-7=2$. There was also confusion as regards order, as the child clearly meant to say 7 from 9 and not 9 from 7. Miles (pers. comm.) also suggests that dyslexics have fewer 'number facts' at their disposal, i.e. being able to recognize 8×7 immediately without further working out.

Other common errors that we have observed include forgetting to carry across, or working from left to right instead of from right to left. This would mean adding the 10s first and then the units, which results in considerable confusion and errors. The problem arises because dyslexics have learned that written language is read from left to right, and then become confused because maths tends to work from right to left. These basic confusions and uncertainties about subtraction and addition often contrast markedly with understanding mathematical concepts. Dyslexics also have problems remembering which particular symbol is used for addition or subtraction, and they often transpose numerals, e.g. writing down 59 when they mean 95.

Tables seem to be a particular problem. Dyslexics often lose their place in the middle of reciting a table, continuing on from a faulty premise, e.g. $2 \times 8 = 15$, $3 \times 8 = 23$, and so on. Children often change to a different table, or use other organizational skills such as using their fingers. Although I am not arguing that these errors apply only to dyslexics – many non-dyslexic children also make similar errors – it does seem that dyslexics make these errors more frequently.

As well as the above, there are more subtle areas with respect to the 'language' of mathematics itself. This is in terms of the way in which things are said, e.g. as regards time, one might say 10 minutes past 5, which is normally written as 5:10. A dyslexic might write this as 10:5, because this is the way in which it is said. In addition, there are many different ways of presenting a simple subtraction task. It may be presented vertically,

$$\begin{array}{r} 24 \\ 16- \end{array}$$

or horizontally, 24 – 16. There are also different ways in which this task can be taught, e.g. decomposition: 4 – 6 'won't go', so take 1 from 2 on the 10s and cross out the 2 to give 1, transfer this across to 14 and do 14 – 6 = 8, and 1 – 1 = 0. On the other hand, one might 'borrow' a 1 from the 2 (10s) in order to give 14 – 6, and then 'pay back' 1 to the bottom of the 10s and there 2 – 2 = 0. Or, one can borrow a 10 from the 10s side and do 10 – 6 = 4, 4 + 4 = 8 and pay back the 1 at the bottom again and 2 – 2 = 0. These different ways of doing the same simple task make it very confusing for children, especially dyslexics who have problems with language and understanding the operations involved.

Dyslexics also have problems with directionality. When measuring angles, dyslexics are often unsure which direction to move from and where to start on the page, and they often lose their place during long operations, such as multiplication. Dyslexics are also often confused by the concepts 'backwards' and 'forwards' when reciting numbers, e.g. is counting 2, 4, 6, 8 counting backwards, forwards, upwards or whatever? (see Durkin & Shire, this volume.) Children sometimes go to great lengths with their own strategies. For example, Joffe (1981) notes an intelligent 14-year-old who, when asked to work out 580 × 7, laboriously jotted down 7 groups of 580 dots, and still got the answer wrong! This indicates that the child had no real understanding of multiplication or basic arithmetic.

Table 16.1 presents a summary of the problems encountered by dyslexics.

Some research findings

There has been very little formal research into dyslexia and maths. Miles (1983) found that dyslexics were more likely to produce the above errors than controls. In an attempt to relate the language difficulties experienced by dyslexics to mathematics problems, Joffe (1981, 1983) has drawn attention to the parallels between written language and number. In both instances, invented symbols are used to manipulate concepts. These symbols are often taught linguistically, and an analysis of error types such as decimal value and regrouping problems, suggests that there may be verbal labelling difficulties. For example, to do the sum 19 + 36, one has to relabel the 9 and 6 as 5 units + 1 ten and then remember to add the 10 unit. Joffe found that dylexics produced 40% more of this type of error than controls, which, she argued, suggests verbal encoding or sound coding difficulties relating to the language difficulties described at the beginning of this

Table 16.1 Problems facing the dyslexic in mathematics

Memory
1 Retention of tables
2 Remembering procedures and rules
3 Retention of information while calculating (e.g. regrouping or addends)

Reading Reading the question, comprehension in problem solving, undertaking required operations
1 Use of ruler, protractor and compass
2 Cutting out
3 Formation of numbers and letters

Direction
1 Horizontal sequencing, i.e. right to left
2 Vertical sequencing, i.e. bottom to top
3 Confusion with $\times$ and $+$ symbols, e.g. making $3 \times 2 = 5$ (i.e. $3 + 2$), or confusion with $-$ and $\div$ symbols.

Organization
1 Sub-skills of the hierarchical system
2 Order of procedures during calculations

Perception
1 Concept labelling, i.e. symbols, names
2 Complex relationships – place value, e.g. 3 and 8 can be used for 38, 3.8 or 380. Also 710 can be confused with OIL, or 21 with 12
3 Conservation is often a slow process

chapter. She also argued that there was poor generalization from specific sums to arithmetical concepts. The term arithmetical is used deliberately here to distinguish it from mathematics in the wider sense. Miles and Ellis (1981) pinpointed what they called 'lexical' encoding, suggesting that losing one's place in tables reflects problems in labelling the relationships between the units or having no 'hook' into the sequence of numbers.

Our own studies have pinpointed difficulties in the way in which maths is sometimes set. Specific problems are often couched in quite difficult language. At a basic level, children have to read the problem (see De Corte and Verschaffel, this volume). However, more subtly, they have to work out what mathematical operation is required, which can involve quite difficult linguistic conceptualizations. For example, 'Patrick went 340 metres on his bicycle, while his friend Gene went 128 metres. If they started together, how far ahead was Patrick at the finish?' This basic sum is very simple and was well within the grasp of the 11-year-olds to whom it was given. However, the majority of them were unable to read the problem in the first place, given the quite difficult wording that was used. Once the problem was read to them, they became confused about small details. For example, they wondered what starting together might mean in terms of the time they took: this they thought might affect the distance that was travelled. They were also confused with the phrase 'being ahead': the children often mixed up the two boys and thought that Gene in fact 'was ahead'. Additional confusion arose over 'Patrick at the finish': some children assumed that 'finish' meant that Gene had ridden for 340 minutes in order to reach the same finish as Patrick.

These are typical examples of focusing on irrelevant details and understanding linguistic concepts. Another example can be shown using a 'trick' question: 'Two boys went around a games field; one went round to the west and one went round to the east. The first boy took one hour and the second boy took 60 minutes. Who took the longest?' The dyslexic children tended to focus on such details as 'It could have been raining and the boy going west was going into the rain' or 'it was uphill going east'. Some simply confused the concepts 'east' and 'west'. As De Corte and Verschaffel (this volume) and Clarkson (this volume) have shown, many children experience difficulties with word problems in mathematics. However, for dyslexic children, the language problems are particularly acute.

Given appropriate help there is no reason why dyslexics cannot overcome their learning difficulties in mathematics. Figure 16.1 plots some of the data we collected using the *France Profile of Mathematical Skills*, which gives age scores for each skill tested. The children were divided into three groups: group 1 was tested upon arrival at school; group 2 had been at school for 1 year; and group 3 had been at school for over 3 years. The data collected showed deviations from the predicted age scores.

It can be seen that when the children first arrived at school, they had problems with all of the basic operations, most especially with multiplication and division. However, they did not seem to have the same weaknesses with regard to measurements, the use of money or simple geometry. Although addition and subtraction began to improve after 1 year, multiplication and division were still causing severe problems.

By the time the children had spent 3 years at the school, they had more or less overcome any difficulties with addition and subtraction, and indeed had moved ahead of their age group. However, because all of the children involved were of above-average intelligence, this was to be expected. Multiplication and division continued to present difficulties.

Practical implications

Despite the paucity of research in this area, the following are some principles and techniques based on teaching experience and on the general principles that are required for teaching dyslexics written language.

1 *The need for a fully structured teaching programme*. These follow the general principles that are required for teaching dyslexics written language. The first is that one should start with basics and build up slowly to more complicated

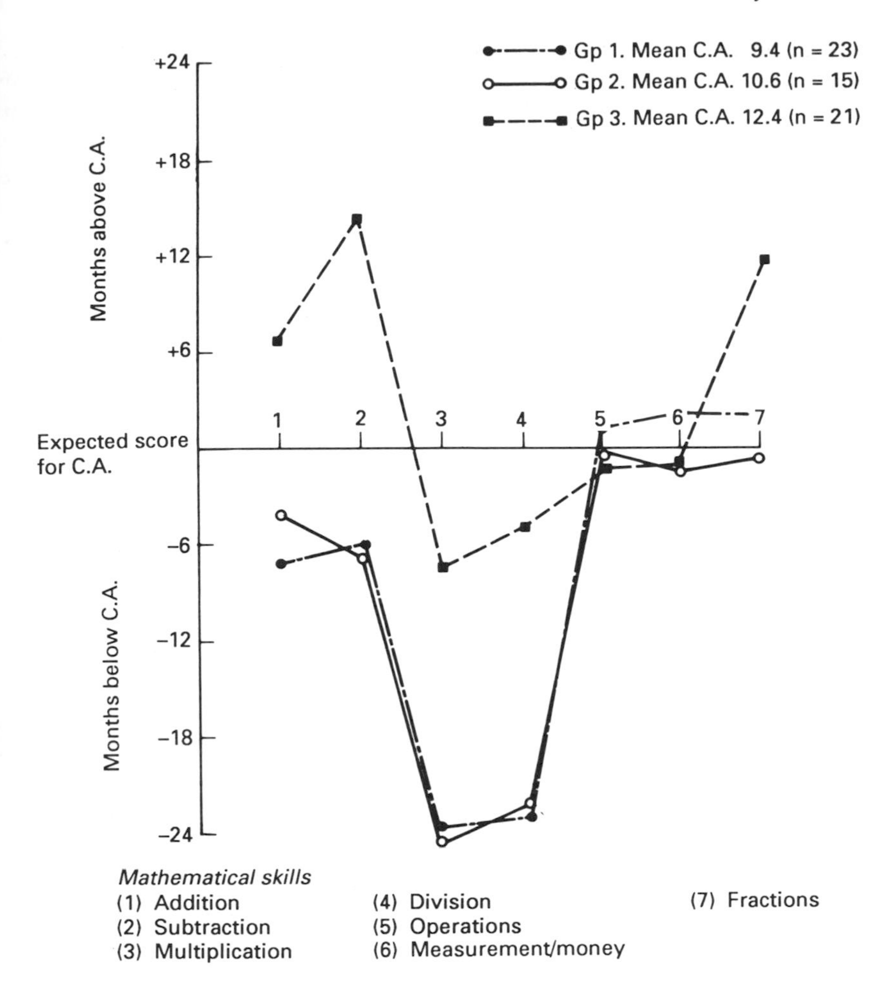

Fig. 16.1 Deviation from age scores of dyslexics on the *France Profile of Mathematical Skills*.

concepts. This is crucial if the dyslexic is to understand mathematical concepts before progressing to more difficult operations. Because a good deal of time is required to be spent on the four basic mathematical operations, progress will be slow.

2 *Expectations must be realistic*. We find it is very important to go over the basics when developing concepts. 'Over-learning' is a key principle for dyslexics; owing to weaknesses in short-term memory, material needs to be presented again and again. To overcome boredom, teachers need to vary the ways in which the material is presented, e.g. by using your computer games, concrete aids, etc. A systematic and carefully thought out syllabus needs to be followed.

3 *Integrated approach to teaching tables*. Even if one were to spend the whole

time teaching tables to dyslexic children, at the end of a 3- or 4-year period, they might still not be proficient at doing them. It would be counterproductive to do this, and we suggest teachers employ table squares and calculators. However, we do reinforce tables daily, using a method similar to 'simultaneous oral spelling'. This is a multi-sensory technique used in the teaching of spelling to dyslexics, which involves relating sound (auditory) to the written (visual) symbol, and then to the touch (kinesthetic) involved in writing. As far as tables are concerned, this involves taking a particular times table, pairing the children off, and then allowing them to test each other. One child recites the table without giving the answers, and the other child then repeats the table but at the same time gives the answers. If correct, the child writes the numbers down, speaking them as he or she does so. When they have been written down, they are read again. This process therefore involves listening, speaking, writing and looking.

4 *Providing success in areas of strength.* With regard to maths, one should try to use the logical, analytical ability of many dyslexics, particularly the more intelligent ones. For example, Dienes Logic Blocks can be used, or when doing a sum such as 23 × 14, one can refer to 14 boys getting 23p each. This simple problem can be divided into 10 boys with 23p each, added to 4 boys with 23p each. Breaking sums down into their logical components is particularly helpful.

5 *Circumventing difficulties.* We find it is best to 'concretize' problems. This makes them much more life-like, e.g. one can refer to 3 × 2 as three bags with two sweets in each. Arms and legs can be used for 2 times, the wheels on a tricycle for 3 times, the wheels on a car or the sides of a square for 4 times, etc. When confronted with fractions, such as 'what is a $\frac{1}{4}$ of 8?', a cake with 8 cherries can be divided equally into 4 pieces, thus giving 2 cherries for each slice. (A real cake can even be baked.)

Using money also helps, e.g. dyslexic children can use 5p and 10p pieces when asked how many 5s there are in 10. Or, they can share 10 apples between 5 boys. Sequential problems can also be focused on by using large cards upon which 100s, 10s and units are written. These can be used to determine the sequential components that are required when adding and subtracting, for example. Given that children seem to be better at working with money, it is useful to relate decimals to our monetary system.

6 *Rewarding correct responses.* This principle should be common to all good teaching, but it is particularly advisable when working with dyslexic children whose self-esteem has often suffered as a result of their disability and who constantly need positive feedback. It is also important to provide such feedback as soon as possible so as to avoid the internalization of incorrect procedures.

8 *Undertaking small sequential steps and recognizing that there are many routes to the same objective.* Because dyslexics have a poor short-term memory, the provision of memory aids and mnemonics is essential, and teachers should be more systematic regarding their teaching layout and presentation skills. Dyslexics also need guidance in how to lay sums out on the page, the sequence of a particular sum, and the formation of numbers and the lines required for basic arithmetic.

Conclusions

Mathematics presents verbal challenges to all pupils, but they are exacerbated among children who experience specific difficulties in reading and writing. Dyslexic children are uncomfortable when working with print, and mathematics relies heavily upon the ability to understand and use written symbols. As a result, dyslexic children – otherwise of normal or even above-average general cognitive ability – frequently fall behind their peers in some areas of mathematics. There is an urgent need for more detailed research into the problems they face and the best ways to help tackle them. However, there is also a basis for teaching mathematics to dyslexic pupils, and I hope that this chapter has indicated some of the practical steps that can be undertaken towards improving the dyslexic child's prospects in this area of the curriculum.

Acknowledgement

I am indebted to Ms Lynne Swan, teacher of mathematics at East Court School, for some of the specific examples cited in this chapter.

Further reading

Snowling, M.J. (1987). *Dyslexia: A cognitive developmental perspective*. Oxford: Blackwell.

Thomson, M.E. (1990). *Developmental dyslexia: Its nature, assessment and remediation*, 2nd edn. London: Cole and Whurr.

17 Teaching maths to young children with language disorders

Chris Donlan and Ella Hutt

We might expect that children with impaired language, usually described as having intact non-verbal abilities, would be inhibited in their counting only by a deficit in number-word acquisition. We might at least conjecture that the other principles of counting would present problems only as a result of linguistic deficit. But experience suggests otherwise. A child with very severe expressive and receptive language impairment may learn to count reliably at a relatively early age, using written or manually signed numerals. In contrast, a child with some expressive ability may have a stable string of number words, may be able to perform certain sorts of 'sums', but may show a persistent and undermining difficulty in the co-ordination of counting principles (see the examples below). Language impairment is itself so diverse (cf. Leonard, 1979) that any simple equation of linguistic deficit with a particular pattern of mathematical learning is likely to be misleading.

Relatively little research is available which has investigated the mathematical progress of language-impaired children, although there is ample evidence that such children do have many problems in the mathematical area. In this chapter, we illustrate some of these problems with reference to specific case studies we have conducted and we suggest several practical steps that might be taken by teachers working with language-impaired pupils.

There is every reason to reject the idea of language disorder as a uniform effect. Bishop (1987) has usefully summarized the complex causation of comprehension problems. It is not easy to separate difficulties in auditory perception, memory or attention, from weak vocabulary, immature grammatical development, or lack of general knowledge about the world or about how language is used. Practitioners will readily testify to extensive individual variation in the nature and effects of impaired language. And although research has yet to evaluate experimentally its effects on mathematical learning, descriptive work by experienced teachers (Hutt, 1986; Grauberg, 1985) suggests that these effects are neither limited to the purely linguistic aspects of mathematical learning nor predictable from conventional profiles of individual linguistic deficit.

In order to teach maths effectively to children with language disorders, then, we must expect a broad variation in performance: we must anticipate difficulties in the non-verbal aspects of mathematical learning at least as great as those experienced by many children whose language acquisition is normal, and we must also be prepared for the child whose mathematical learning is a relative strength.

Three examples will serve to illustrate the breadth of the problem. After outlining the problems of three unique individuals, we shall go on to propose detailed teaching strategies which we believe will be of general value.

Case study 1: Difficulties with the number system and place value

Susie is an 8-year-old of average intelligence. She has been placed in a special language school for 18 months but is soon to return to the mainstream. Her comprehension of language shows a narrow deficit as measured on the Reynell Developmental Language Scales (age level $6\frac{1}{2}$–7 years at C.A. 7:7). She has an immature phonological system which limits intelligibility; her syntax is improving, she uses a variety of sentence structures but still shows immature past-tense forms. Her progress with written language is starting to accelerate but she still has difficulty in sequencing the letters within words. She uses phonic strategies in preference to a whole-word approach to spelling.

Susie has spent the greater part of a 45-minute maths session working to produce a number sequence on a long strip of paper. She has seen others doing this and desperately wants to succeed herself. Eventually, she brings her work to the teacher, proud of what she has produced but frustrated that she 'can't get to a hundred'. The sequence she produced was continuous. It is reproduced here in sections (Fig. 17.1).

Susie's number sequence amply illustrates the resourceful way in which children use their knowledge and intelligence to produce an approximation to adult performance/requirements. Their efforts provide us with tangible and fascinating evidence of the way their thoughts are developing. There are many useful observations to be made from Susie's sequence:

1 The initial point of breakdown in the sequence is the very point at which the verbal number sequence breaks into the 'teens, and where the order of the written numerals seems to contradict the spoken. Susie clearly manifests the decade problem discussed by Fuson (this volume). 'Thirteen' might plausibly suggest 3, 10 and therefore 30, especially to the young learner who may be attempting to synchronize her writing with verbal rehearsal of the sequence. 'Thirteen' and 'thirty' present auditory confusion in any case.
2 Susie's 1–9 sequence is consistent within tens and between adjacent 'ten-markers' (30, 40, 50, 60 in line 2; and 40, 50 in line 4).
3 The frequent but inconsistent reversal of numerals 2, 3, 4, 5, 7 does not seem to affect the sequence, but the formation of 9 presents more complex problems. Its formation varies in adult script and printed text. It may have been a visual link that provided Susie with the transition from 69 to 30 in line 3.

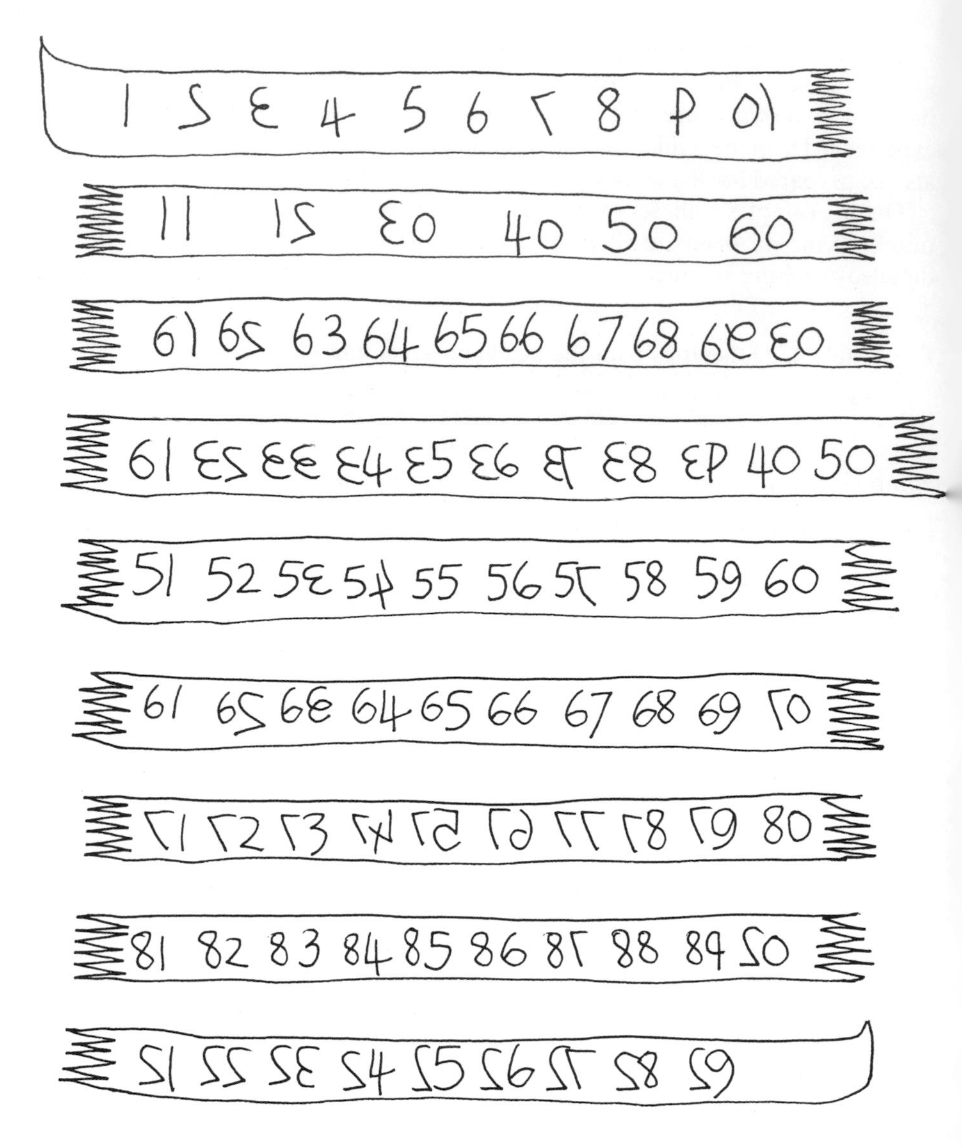

Fig. 17.1. A written number sequence produced by Susie, aged 8, attending a school for language-impaired children.

4 If verbal rehearsal was a significant strategy (as seems most likely), then it might explain the transition 89, 20 which might be suggested by the spoken sequence 'eighteen, nineteen, twenty'.

There is as much to learn from the consistencies of Susie's number sequence as there is from its inadequacies.

Case study 2: Problems with counting

Tom is 8 years of age. He has recently been transferred to a language unit following MLD placement. Age-appropriate (or better) scores on tests of language comprehension, confirmed by informal findings for functional comprehension, gives an impression of average intelligence. His scores on some non-verbal subtests from Intelligence Scales, however, as well as performance on Raven's Progressive Matrices, are startlingly low (10th centile or lower). His expressive language is both phonologically and grammatically limited.

According to reports, Tom has entered the unit able to manage addition and subtraction within ten. Accordingly, he is working his way through the early addition section of a published maths scheme. His workbook offers him various sorts of visual aid to counting the total. All seems well: then, inexplicably, he breaks down in tears. In an attempt to find out what is wrong, part of a specialized assessment procedure is used (MAP: Donlan and Hutt, 1989).

MAP, a Mathematics Assessment Procedure for young language-impaired children, provides a profile of mathematical learning which separates component skills, looking in particular at the details of counting. While Tom was able to subitize (i.e. recognize/quantify quickly and accurately, without overt counting) dice-type patterns up to six, he made errors when counting objects or pictures of six objects, consistently failing to co-ordinate his tagging of the objects with the spoken number sequence. This immaturity was especially surprising in view of Tom's ability to retain in his mind any totals reached (an aspect of the so-called Cardinality Principle which often emerges after co-ordination is mastered: cf. Gelman and Gallistel, 1978; Fuson, this volume). Furthermore, Tom was able to count-on from 3 and 4, given objects in view, more accurately than he counted from 1. He was also able to produce a verbal number sequence to 20, and to read, write and sequence the numerals 1–10.

The stress which certain counting tasks produced in Tom made it fairly clear that his problem with the addition sums had been at this level. His teacher later observed that he showed a preference for a particular rigid counting-up (rather than counting-on) strategy to perform sums, using Unifix to represent both addends of a sum and then counting them up together. It seems likely that the visual confusion presented by some of the workbook formats had overloaded Tom's counting ability. It was decided to concentrate on strengthening his counting skills before going further.

Case study 3: The challenge of time concepts

John is 7-years-old. Three years ago he lost the facility for speech and for understanding spoken language, though up to that time he had been developing normally. His comprehension is now virtually nil. Although his hearing from non-speech sounds is normal, he appears not to process spoken language at all, often failing to respond to his own name. He has no expressive vocabulary, communicating his needs through gestures combined with a very limited repertoire of sounds. John has recently started to learn some signs in Paget-Gorman Signed Speech. He is placed in a language unit from which his limited integration is carefully monitored.

John enjoys number work. Given regularity of format, using simple workcards provided by his specialist teacher, and avoiding the potential confusion of published schemes, John is able to deal flexibly with the sequence of numbers 1–20, he can manage simple computation, and outperforms some of his peers whose language impairments are less severe.

Though John's non-verbal performance on categorization and grading/seriation tasks shows relatively advanced reasoning ability, he has no understanding of words like 'all', 'some', 'another', 'first' or 'next'. In the same way, his counting skills and knowledge of the number system are relatively intact, but when required to use the most basic of these skills in a simple practical task (counting small amounts of 1p pieces), he is confused by the verbal labels involved in the instruction. His understanding of the language of size extends no further than 'big' and 'little'. The severity of John's language impairment has not prevented the development of mathematical thinking and numeracy in the strictest sense, but it crucially affects their practical application and further development.

Concepts, especially abstract concepts, have to be specifically taught to children like John. Time concepts are highly abstract but also of crucial importance for communication.

Practical implications

Teaching the number system 1–100 and place value

Special attention must be given to the discrepancies between the spoken number words ('one', 'two', 'three', etc.) and the numerals (1, 2, 3, etc.). Though we as adults use them interchangeably and automatically, we should be aware that the spoken number sequence is significantly different from the more regularly 10-based numeral system. Verbally, we use unique forms for numbers 'one' to 'twelve' but we only use unique numerals 1–9; we shift the place of the first numeral and follow it with a zero to signify 'ten'. The verbal system does not change until it reaches 'thirteen'. Unique forms now give way to a word that presumes to represent 'ten' and 'three' but does so in reverse order.

For number novices, who probably count to themselves (verbally) as they write down the numbers (see Sinclair, this volume), the word 'thirteen' might logically seem to be represented by 30 or 31. This sort of confusion must be especially difficult for children whose auditory perception and/or sound production fails to discriminate between -*teens* and -*ties*.

Supportive teaching should be guided by two principles which are common features of work with language-impaired children:

1 Use systematic visual patterns to aid learning.

2 Give separate consideration to the auditory/verbal component of any task, and structure its requirements to suit the needs of particular learners.

Systematic visual patterns for the numbers 1–100 can be provided in many forms. The numerals themselves presented in a 10 × 10 '100 square' show many memorable features which can be enhanced with colour coding. Hutt (1986) describes and illustrates a 'closed 100 square' which can be used in a variety of games or simply as a point of reference. Grauberg's (1985) 'Paper Abacus' allows the construction of quickly recognizable 'number pictures' representing numerals up to 100.

Both systems have the advantage of locating each number within an organized pattern which is reliable and predictable. This approach differs significantly from the more traditional illustration of particular isolated numbers as bundles of tens and remaining single items. An illustration of this sort stresses the 'cardinal' value of the number concerned (the total reached when counting a set of this size) rather than its 'ordinal' location (the position of a certain number in the number sequence). Both cardinal and ordinal aspects of the number systems should be taught, but it is important that the teacher understands the distinction and avoids a confusion of aims. Cardinal value is directly related to counting and so is best introduced with concrete items or structured materials for counting, and not by static illustration on a printed page; as Sinclair (this volume) points out, many pupils in the early stages of maths work do not grasp the cardinal value of written numbers in isolation. Ordinal location refers to position within a sequence and so is best introduced within a static, organized framework.

In both cardinal and ordinal work, it is advisable to stress the identification of written numerals with number words where they are most compatible, i.e. the sixties, seventies, eighties and nineties. There is no harm in directing the child's attention to these regularities at an early stage. It helps to develop confidence. Also, it is easier for a child to grasp that groups as well as ones can be counted when there are more groups, for example six tens, than when there are fewer, for example only three tens. In Susie's case (see above), the 'ordinal' system became established through a confident repetition of the 1–9 sequence, especially where the spoken and written numerals corresponded best. She came to grips with the ordinal system before fully understanding its cardinal values. Perhaps it was necessary for her to clarify the awkward sound/symbol confusions before attributing full numerical meaning to them.

Understanding how auditory perception develops and how this development is related to a child's growing ability to produce complex and finely differentiated sounds is a specialized field of study in itself. Wherever possible a speech therapist's advice should be sought to clarify general principles as well as to assess particular children's strengths and weaknesses in the absence of specialist help, it is clearly better to underestimate rather than overestimate children's abilities. The priority given in the previous paragraph to the more systematic sound/symbol correspondence of the number range 60–99 is especially important for learners whose language is impaired. Though it may be tempting to isolate the pairs of numbers which cause most trouble (13/30, 50/15) and attempt to teach the difference, this is likely to overload and confuse children whose auditory discrimination is weak. It is far better to work on the sixties first, to become confident of the distinctions within a narrow range, and to stress the sound/symbol correspondence at its strongest. An infinite number of pair-matching games with a lotto-type auditory/verbal component can be constructed around the particular number range required.

Teaching counting

Gelman and Gallistel (1979) identify the basic principles which are usually established around the age of 5–6 years, and Fuson (this volume) stresses the importance of counting as the foundation of development in number. Especially for children with learning difficulties, including language difficulties, there may be a surprising imbalance in counting behaviours (cf. Tom, above).

The nature and context of any counting task radically alters counting behaviour. The most obvious influence is that of set size. A child who is well able to co-ordinate tagging and partitioning for sets of four or five may completely lose this skill when presented with six or seven items. Of equal importance is the complexity of the array. A set of six identical toys in an evenly spaced line might allow a child to count confidently and accurately, but if the set to be counted is presented in picture form, perhaps a group of six cars of different colours and presenting different or incomplete outlines, then the task is vastly more difficult (see Potter and Levy, 1968).

The examples provided so far require the simple enumeration of given sets. But counting skills are frequently employed in contexts which are far less clear-cut. If a child is asked to get seven cups from a pile containing a greater number, then a sophisticated grasp of the cardinal principle is required for execution of the task. If the task is to find five cars in a box of different toys then further confusion is possible, not only through the introduction of a sorting task but also through the increased demands on verbal comprehension and memory; all provide potential interference. Control of the context of counting may seem no more than a commonsense requirement, but it is easily overlooked in practice. For the child with impaired language a structured approach is essential for building solid counting skills in association with secure verbal comprehension and memory.

A broad assessment of counting skills provides a firm base for structured teaching. The Maths Assessment Procedure for young language-impaired children includes a progression of counting tasks of increasing complexity. At the early stages, an important comparison is made between counting *per se* and the related skill of subitizing. Subitizing is a useful and practical skill which adds flexibility to counting behaviour and which can improve rapidly with practice. It has a strong visual component appealing to children with limited auditory/verbal processing ability. MAP compares performance on simple counting, subitizing and complex counting tasks in different number ranges and includes checks for understanding the cardinal principle (Table 17.1).

Table 17.1 Progression of counting tasks (from Donlan and Hutt, 1989)

	MAP counting/quantifying		
	Task	*Materials*	*Arrangement*
Part 1	Count/subitize	Identical toys 1–6	Simple
	Count/subitize	Dots 1–6	Dice-type; other
	Match	Pictures 1–6	Complex
	Sort and count	Toy cars	Selected from mixed toys
Part 2	Share	Pennies	Share between 6
	Count	Pictures 6–9	Complex
	Count-on	From 6 objects	All in view
Part 3	Take away	From 6 objects	In view
	Count-on	Up to 11 objects	First quantity hidden

An essential part of an assessment of this sort is observation and recording of the quality of the child's response. It is not enough simply to note a failure to count five objects accurately; was there a failure in the number string, in the co-ordination of tagging and partitioning, or through interference from the context of the task?

By following the progressive mastery of counting, relating it to verbal comprehension and memory, the teacher can approach with excitement and confidence the transition from the establishment of the cardinal principle through counting-on to real addition.

Teaching time concepts

A child like John (see above) is unaware of many concepts which are acquired with comparative ease by most children. Such concepts have to be specifically taught to children with severe disorders of language. Probably the most difficult global concept is that of the passing of time. Awareness has to be increased gradually, by ensuring that vocabulary is provided which helps to define accurately the child's growing understanding of periods of time. A series of simple visual aids has been found to be effective.

First, each day must be given its name. The recurrence of these names in a seven-day cycle can be shown on a week-clock with one hand, which is moved forward daily to the name of the new day.

Next, come two important broad divisions within a day: morning and afternoon. And, does John realize that, although he goes to bed at seven o'clock and gets up at seven o'clock, a long period of sleeping-time separates the two? A 24-hour card clock can represent this: small drawings are used to indicate routine daytime activities, the 'morning' and 'afternoon' areas being coloured differently; and a black area surrounding a bed indicates night-time. The two meanings of 'day' must also be discussed, i.e. the time between one midnight and the following one, and the hours of daylight.

Then comes the sequence of the days themselves, and the generalizations of the adverbs 'today', 'yesterday' and 'tomorrow'. A daily ritual is established, changing the days on a three-day calendar:

Monday was yesterday.
Tuesday is today.
Wednesday will be tomorrow.

When a child is ready for the dates themselves to be entered, a seven-pronged quarter-circle can indicate a week's days, matched to dates within a month. An outer three-pronged segment, showing 'yesterday', 'today' and 'tomorrow' is moved on daily. If a conventionally shaped calendar is used, it is advisable not to present all the dates at once, but to add the date in the appropriate space each morning. In this way a salient visual impact is made. A birthday can be marked in advance with the child's name.

The passage of clock-time within a school day is represented by a chart of two vertical columns of clocks, each of which shows the position of the hour hand only, at the times at which the school bell is rung. The spaces between the clocks show clear symbols of activities such as painting or reading. Whenever a bell is heard, a child moves a marker labelled 'now' to its place opposite the relevant activity.

Time-telling must not be taught on a digital clock. This would wreck understanding of the decimal system of notation. What else proceeds '158, 159, 200'! So much mathematical understanding must precede each stage of learning to 'tell' the time that it is essential to teach this skill slowly: the o'clock times, then 'half past' times, both with the hour hand only. Each child must learn to read the proportional distance between consecutive numerals of the hour hand. After 'quarter past' and 'quarter to', five-minute intervals can be attempted, but not until the products of one, two, three, four and five fives are known. Then the minutes between the multiples of five provide good practice in counting-on.

Some attempt is made to provide a visual representation of the four complete weeks in each month. And, finally, the ordinal number of the current year, having removed the number of the century, is shown as one of a series of ten or twenty years. One strip per child matched against the master strip gives some visual indication of the relationship between the current year number, the year of the child's birth, and his or her age.

Conclusions

We hope that this chapter provides the reader with some idea of the wide variation shown by language-impaired children in their learning of mathematics. We have certainly not exhausted the range of difficulties that such children manifest, but we have attempted to illustrate the bases for observations that teachers can make of their pupils' problems. Having invested time in such observations the teacher is well prepared for structuring a learning programme, and we have suggested specific strategies and a content for a series of programmes. Finally, we hope we have conveyed something of the fascination and enjoyment in store for the language-impaired child and the teacher who allow themselves to have fun with maths.

Further reading

Donlan, C. (1989). Patterns of mathematical learning associated with language disorder. In K. Mogford and S. Sadler, (eds), *Child language disability*. Clevedon: Multilingual Matters.

Donlan, C. and Hutt, E. (1989). *M.A.P.: Mathematics Assessment Procedure for young language-impaired children*. London: I CAN.

Section 7

Cross-linguistic issues

18 Chinese-based regular and European irregular systems of number words: The disadvantages for English-speaking children*

Karen C. Fuson and Youngshim Kwon

Chinese and some other Asian languages (e.g. Burmese, Japanese, Korean and Thai) have regular named-value systems of number words in which a number word is said and then the value of that number word is named (five *thousand* seven *hundred* two *ten* six). Many European languages have regular named-value number-word systems for the values of 100 and 1000, but they are irregular in different ways below 100 (see Table 18.1). These irregularities have serious consequences which affect children's numerical learning adversely in several different ways. Sufficient research is available to describe these consequences for English-speaking children (mostly in the USA), and therefore the focus in this chapter is on the English language. However, many of the consequences discussed here would seem to apply to the other irregular European number-word systems, though some details might vary. This chapter will briefly review English-speaking children's relative difficulties compared to Asian children *in learning the number-word sequence, in adding and subtracting numbers with a sum between 10 and 18, in constructing adequate mental representations for multidigit numbers*, and *in adding and subtracting multidigit numbers accurately and/or meaningfully*. Making linguistic comparisons is complex because most such comparisons also involve many non-linguistic cultural factors that might affect learning. This problem is reduced in this case because the linguistic effects are supported by data concerning different kinds of errors made or different solution procedures used in the regular and irregular languages, rather than a simpler accelerated learning in one language that might be due to more general cultural factors.

*Parts of this chapter were presented at the Biennial Meeting of the Society for Research in Child Development, Kansas City, April 1989.

Table 18.1 Number words

	French	*Spanish*	*Italian*	*German*
1	un, une	uno, una	uno, una	eins
2	deux	dos	due	zwei
3	trois	tres	tre	drei
4	quatre	cuatro	quattro	vier
5	cinq	cinco	cinque	fünf
6	six	seis	sei	sechs
7	sept	siete	sette	sieben
8	huit	ocho	otto	acht
9	neuf	neuve	nove	neun
10	dix	diez	dieci	zehn
11	onze	once	undici	elf
12	douze	doce	dodici	zwölf
13	treize	trece	tredici	dreizehn
14	quatorze	catorce	quattordici	vierzehn
15	quinze	quince	quindici	fünfzehn
16	seize	dieciséis	sedici	sechzehn
17	dixsept	diecisiete	diciassette	siebzehn
18	dixhuit	dieciocho	diciotto	achtzehn
19	dixneuf	diecinueve	diciannove	neunzehn
20	vingt	veinte	venti	zwanzig
21	vingt et un	veintiuno	ventuno	einundzwanzig
22	vingt-deux	veintidós	ventidue	zweiundzwanzig
23	vingt-trois	veintitrés	ventitre	dreiundzwanzig
24	vingt-quatre	veinticuatro	ventiquattro	vierundzwanzig
25	vingt-cinq	veinticinco	venticinque	fünfundzwanzig
26	vingt-six	veintiséis	ventisei	sechsundzwanzig
27	vingt-sept	veintisiete	ventisette	siebenundzwanzig
28	vingt-huit	veintiocho	ventotto	achtundzwanzig
29	vingt-neuf	veintinueve	ventonove	neunundzwanzig
30	trente	treinta	trenta	drei-ssig
31	trente et un	treinta y uno	trentuno	einunddreissig
39	trente neuf	treinta y nueve	trentnove	neununddreissig
40	quarante	cuarenta	quaranta	vierzig
50	cinquante	cincuenta	cinquanta	fünfzig
60	soixante	sesenta	sessanta	sechzig
70	soixante-dix	setenta	settanta	siebzig
80	quatre-vingt	ochenta	ottanta	achtzig
90	quatre-vingt-dix	noventa	novanta	neunzig
99	quatre-vingt-dix-neuf	noventa y nueve	novantanove	neunundneunzig
100	cent	cien	cento	hundert
101	cent et un	ciento uno	centouno	hunderteins
125	cent vingt cinq	ciento veinticinco	centoventicinque	hundertfünfundzw
4313	quatremilletrois-centstreize	cuatromiltres-cientostrece	quattromille-trecentotredici	viertausenddreihun dreizehn

Notes: The words for 4313 are one word in the first four languages; they are hyphenated here for lack of space. The Chinese words are given in English to show their structure. Burmese, Japanese, Korean, Thai, and Vietnamese number words have the same regular structure as the

ish	*Chinese*	*Positional base-ten*
	one	one
	two	two
e	three	three
	four	four
	five	five
	six	six
n	seven	seven
t	eight	eight
	nine	nine
	ten	one zero
en	ten one	one one
ve	ten two	one two
een	ten three	one three
teen	ten four	one four
en	ten five	one five
en	ten six	one six
nteen	ten seven	one seven
teen	ten eight	one eight
teen	ten nine	one nine
nty	two ten	two zero
nty-one	two ten one	two one
nty-two	two ten two	two two
nty-three	two ten three	two three
nty-four	two ten four	two four
nty-five	two ten five	two five
nty-six	two ten six	two six
nty-seven	two ten seven	two seven
nty-eight	two ten eight	two eight
nty-nine	two ten nine	two nine
ty	three ten	three zero
ty-one	three ten one	three one
ty-nine	three ten nine	three nine
ty	four ten	four zero
y	five ten	five zero
y	six ten	six zero
enty	seven ten	seven zero
hty	eight ten	eight zero
ety	nine ten	nine zero
ety-nine	nine ten nine	nine nine
e hundred	one hundred	one zero zero
e hundred one	one hundred one	one zero one
e hundred twenty five	one hundred two ten five	one two five
ur thousand three hundred thirteen	four thousand three hundred ten three	four three one three

inese words, and Bahasa (the Indonesian formal language) and Tagalog (the language used in
mary schools in the Philippines) are like Chinese words except they have a teen word and a
v irregularities (Vietnamese also has a few tonal and consonant irregularities).

Relative difficulties in learning the English number-word sequence

The English system of number words does not directly name the ten and one values in two-digit numbers. Several features of English even make it difficult to see this underlying tens and one structure and to see how the first nine numbers are re-used to make the decade words:

1 The existence of the arbitrary number words 'eleven' and 'twelve' that do not indicate their composition as 'ten and one' and 'ten and two'.
2 The irregular pronunciation of 'three' in 'thirteen' and 'five' in 'fifteen' that obfuscate the re-use of the words 'three, four, . . ., eight, nine' with 'teen' to make the 'ten three' to 'ten nine' words.
3 The tens/ones reversal only in the teen words so that the 'four' is said first in the teen word ('fourteen' instead of 'teenfour' or 'ten four') but is said second in all of the other decade words ('twenty four').
4 The irregular pronunciation of the decade words 'twenty', 'thirty' and 'fifty' that mask for many children the relationship of the decade names to the first nine number words.
5 The use of two different modifications of 'ten' (i.e. 'teen' and 'ty') neither of which clearly says 'ten'.

Table 18.1 indicates that other European languages have many of these irregularities, and French, Italian and Spanish also have a reversal and change of form in the middle of the teens (see Menninger, 1969, for more about European number words).

These irregularities in the words between ten and one hundred require children to memorize major parts of the English number-word sequence without seeing patterns other than the one through nine repetition within the decades (Fuson *et al.*, 1982; Siegler and Robinson, 1982) and without seeing units of tens within this sequence. Consequently, children make more errors and more kinds of errors in saying the English sequence than do their peers who are learning the Chinese regular named-value sequence in which tens are explicitly named; errors in English are made in many different places, whereas those in Chinese are concentrated at decade or hundred changes, indicating that Chinese children do see the pattern in their number-word sequence (Agnoli and Zhu, 1989; Miller and Stigler, 1987). Chinese children say their regular words 'ten' through 'ten five' faster than English children say their irregular 'ten' through 'fifteen' words (Agnoli and Zhu, 1989). Other details of difficulties children have learning the English number-word sequence are given by Fuson (this volume).

In general, and not surprisingly, the kinds of errors made in learning a number-word sequence seem to depend upon the structure of the sequence. Deaf preschoolers learning American Sign Language (ASL) from their deaf parents may skip signs that are difficult for them to make with their fingers or may confuse the production rules that generate the number words (Secada, 1985). Italian-speaking children show particular difficulties with the reversal in the upper teens (see Table 18.1); Agnoli and Zhu, 1989).

Relative difficulties in adding and subtracting numbers with a sum between 10 and 18: Construction by English-speaking children of a unitary representation of number words

The lack of an obvious tens and ones structure in English number words between ten and one hundred results in the construction by English-speaking children of a unitary mental representation of number words in which each number word is a single unit. This unitary representation goes through several developmental levels of increasingly efficient and abstract solution procedures for addition and subtraction situations (see Fuson, this volume, for a summary of these levels and Fuson, 1988, for a more detailed discussion of these successive unitary representations). With these unitary representations children do not group objects into tens or count by tens; each number is composed of that number of single units, whether the units are presented by objects or by number words.

This developmental sequence of US children's addition and subtraction solution procedures is almost entirely a secret underground movement within school classrooms. Children invent most of these procedures for themselves without support from their textbooks or teachers. Textbooks merely move from addition and subtraction problems where objects are given with the numbers to problems given in numbers with no available objects (Fuson *et al.*, 1988). Through practice, children are supposed to remember all of the addition and subtraction facts. Because most are unable to remember all of these facts, they move through the above developmental sequence of solution procedures for facts not yet memorized. However, they do so slowly, with a considerable number of second-graders still not at the highest level (Carpenter and Moser, 1983, 1984; Steinberg, 1984). The solution of subtraction problems with sums up to 18 is particularly delayed. Many US textbook series do not even present the most difficult single-digit addition and subtraction problems in the first grade (Fuson *et al.*, 1988).

Asian children learn single-digit sums and differences more rapidly than US children (e.g. Song and Ginsburg, 1987). All such sums and differences are presented in the first grade in China, Japan, Korea and Taiwan (Fuson *et al.*, 1988; Fuson and Kwon, in press). Children in these countries are taught particular methods of adding and subtracting numbers with sums between 11 and 18. These methods all depend on the clear tens and ones composition of these numbers 'ten one' through 'ten eight'. The up-over-ten method is taught for addition: one addend is partitioned into the number that makes ten with the other addend and the left-over number. *Eight* plus *five* is thought of as '*eight* plus two from the *five* is ten plus the three left over from the five is *ten three*'. Some US children also invent this over-ten method (e.g. Carpenter and Moser, 1983, 1984). However, this method is more difficult in English because the 'ten plus x' sums (e.g. ten plus three is thirteen) have to be learned rather than being given automatically by the counting sequence as 'ten three'; many US first- and even second-graders do not know these sums and have to count up from ten to find out how many 'ten plus two' or 'ten plus four' is (e.g. Steinberg, 1984). US children also commonly lack another prerequisite for the over-tens method: there is much less emphasis in the USA than in China or Japan concerning the number that makes ten with a given number (e.g. for eight plus five, one needs to know that eight plus two is ten) and thus many first-graders have to count to find out how many to put with a given number to make ten.

Two different methods are taught for subtraction in Japan, Korea and Taiwan. The 'down-over-ten' method is the reverse of the up-over-ten method: the number being subtracted is split into the number that is over ten, and the rest is then subtracted from ten (ten three – 8: the 8 is split into 3 and 5 – because the three taken from 'ten three' leaves ten – and the 5 is then subtracted from this ten = 5). The 'subtract-all-from-ten' method essentially turns subtraction into an additive procedure: the number being subtracted is taken from ten and the resulting difference is added to the amount over ten (ten three – 8: ten – 8 = 2 plus the three in the 'ten three' = 5).

The use of a unitary representation that has no larger units of ten does not mean that such users cannot learn to read and write two-digit numerals. These written numerals are related to the patterns in the English number-word sequence: the first digit suggests (but does not necessarily specifically name, as in 'twen') the decade name and the second digit tells the word said after the decade name, except for the reversed irregular teen words. Ross (1986) found that all second-graders sampled from a wide range of classrooms were able to count collections of as many as 52 objects and could write the two-digit numeral that told how many objects there were. However, more than half of the second-graders and 15% of the fourth-graders showed no knowledge of tens and ones even as labels for the two digits, indicating that they had only unitary representations for these numbers. The teens reversal in English (and other European languages) does create special problems for writing the numerals for the words between ten and twenty. The words for the ones digits are said first but written second: one says 'fourteen' but writes 14. Thus, many children write 41 for fourteen, following the pattern of the words (some children say that the 1 with the 4 'teens' it, i.e. makes it a teen word). This strategy of writing the ones word wherever it is said works for all the words between twenty and one hundred: one says 'twenty four' and writes 24. So the reversal in the English teen words is particularly troublesome. In Chinese, there is of course no problem with writing any two-digit number or with linking the digits to tens and ones labels: one says 'two ten four' and writes 24 and one says 'ten four' and writes 14.

Relative difficulties in learning adequate representations for multidigit numbers: Interference of the unitary representation when constructing the necessary named-value and positional base-ten representations

The English and Chinese systems of number words are measure named-value systems, and the system of written multidigit number marks used in most countries is a positional base-ten system (see Fuson, in press, for a fuller discussion of the features of these systems and of the different mental representations children construct for these systems, and see Table 18.1 for positional base-ten number words that illustrate some of these features). The numbers in both of these systems are composed of different kinds of multi-units – larger and larger units – and not just of single-unit items as in the unitary number representations. Measure named-value systems have the following features:

1. Each named value (e.g. tens, hundreds, thousands) is a collection of units.
2. New larger values are made by regular ten-for-one trades.
3. Because each new value must be named, large numbers are limited by known names.
4. The value is conserved if named value/digit pairs are out of order.
5. Zero is not needed.
6. Quantities of each value direct multidigit addition and subtraction by adding or subtracting like values.
7. Trading for too many (+) and for not enough (−) in multidigit addition and subtraction are directed by the values of the quantities.
8. Values can have ten or more of that value (e.g. the expression 'fifteen tens' is meaningful).

Positional base-ten systems have the following features:

1. Values are not named but are implicit in the position of a digit relative to the ones position.
2. New larger values are made by regular ten-for-one trades.
3. Because these new larger values just take positions to the left, large numbers are not limited – one can make as large a number as places one can write.
4. The value is not conserved if digits are out of order (e.g. 14 is different from 41).
5. Zero is needed for missing values to keep digits in their correct relative positions.
6. The meaning of positions built up by ten-for-one trades (their values as products of ten) directs multidigit addition and subtraction as adding or subtracting like products of ten (i.e. like positions).
7. Trading for too many (+) and for not enough (−) in multidigit addition and subtraction are directed by the values of the positions.
8. Positions cannot have ten or more in that position.

In order to understand these features of named-value English words and positional base-ten written number marks for large multidigit numbers, children must construct named-value and positional base-ten mental representations for the words and the marks, and relate these representations to each other and to the words and the marks.

English-speaking children show considerable difficulty in constructing multi-unit named-value and positional base-ten representations in contrast to children who speak regular named-value Asian languages. These languages evidently help children construct named-value and base-ten representations for two-digit numbers. Miura (1987) found that Japanese-speaking first-graders living in the

San Francisco area used the tens and units in base-ten blocks to make named-value representations of five numbers between 11 and 42 considerably more than did English-speaking first-graders. The latter instead made unitary count/cardinal representations from single units: 28 little cubes instead of two longs (each long is ten units long) and eight units. Similar results were reported for Chinese, Japanese and Korean children compared with US children (Miura *et al.*, 1988) and even for Japanese first-graders before any work on tens compared to US first-graders after instruction on tens (Miura and Okamoto, 1989). The persistence of this unitary view of number even in US adults is indicated by the difficulty adults had accessing a base-ten rather than a unitary meaning for a two-digit number, even when it would have helped them in a task (Heinrichs *et al.*, 1981).

Because most English-speaking children find sums and differences to 18 using a unitary representation (even the result of a memorized number fact is likely to be in the form of a unitary representation), English-speaking children must shift back and forth between a unitary representation and a multi-unit named-value or positional base-ten representation to add or subtract multidigit numbers meaningfully. This representation shifting, and the difficulty English-speaking children initially have in thinking in tens and ones, is illustrated by the extra step for multidigit addition that was invented by first-graders using base-ten blocks in Fuson (1986). At the beginning of the addition learning, some children would find the sum of a given column, e.g. that seven plus five is twelve, but they would not know how many tens and ones were in twelve because it had only a unitary sequence or cardinal meaning for them (twelve as coming after eleven or as a pile of objects they would get if they counted out twelve things). These children would write down the twelve as 12 using an available rote word-numeral association. They would then look at the 12 and think of it as 'one two' or 'one ten two ones' in order to find that twelve has one ten and two ones; then they would trade the 1 ten and record the 2 in the ones column. Many second-graders in Fuson and Briars (1990) used this extra step on some problems in the post-test. All of these children evidently found the support of the written two-digit marks helpful in shifting from the unitary to the named-value representation that was required to decide what to trade.

It may be that working with four-digit numbers, as the children who invented this step did, facilitates this named-value view of two-digit marks. In contrast, Madell (1985) found that many 6-year-olds using base-ten blocks to invent procedures for adding two-digit numbers held to their unitary mental representation rather than constructing a named-value representation for the blocks: to do $48-14$, they would trade a long for ten units in order to get eighteen units from which they could take fourteen units. They did not use the multi-unit solution supportable by the blocks: think of the 14 as 1 long and 4 unit and take the 1 long from 4 longs (i.e. one ten from four tens) and the 4 units from the 8 units to get 3 longs and 4 units (i.e. 34). Even when English-speaking children get better at knowing the tens and ones in single-digit sums between eleven and eighteen and do not need the support of the written two-digit sum to decide the tens and ones, in most cases they must still switch back and forth conceptually between the unitary representation with which they find the sum of the single-digit numbers in a column and the multi-unit representation that directs their trading when they have too many in a given column.

Asian children speaking a Chinese-based language have linguistic support for the multi-unit representation and do not necessarily even need to shift representations. Their regular named-value words give all sums over ten in a linguistic multi-unit form: seven plus five is 'ten two'. Even if this 'ten two' is initially thought of with a unitary counting/sequence representation with which the sum of seven and five was found, the words themselves suggest what to do with these 'too many' ones. They suggest that the ten in 'ten two' be put with the tens in the other multidigit number and that the 'two' ones be recorded in the ones column. Furthermore, these words are likely to have quantitative 'tens' and 'ones' meaning (and not just be verbal tens and ones labels) because the over-ten method taught in mainland China, Japan, Korea and Taiwan to first-graders for adding sums to 18 supports this quantitative interpretation by using the value of ten in the addition or subtraction procedure (this ten is built up by adding one addend to part of the other addend).

Chinese-based languages clearly are better than English for dealing with situations in which there are more than nine ones because the named-value of ten suggests what to do with the extra ones (i.e. with the ten). When dealing with sums over ten in other columns – situations in which there are more than nine tens, hundreds, thousands, etc. – there are two different approaches that might be taken within a multi-unit named-value representation and within a positional base-ten representation. Again the different language forms support different thinking. A unitary representation within a named-value representation was used by English-speaking children who had used base-ten blocks (Fuson and Briars, 1990) in explaining tens sums that exceeded one hundred: 'That's 8 tens and 8 tens is sixteen tens and ten of those tens makes one hundred and six tens left, so trade the hundred to the hundreds place and write the 6 tens here. It's one hundred and six tens.' With this unitary representation of tens, ten of the tens must be traded for one hundred, i.e. a value trade must occur. In contrast, many Korean children are not permitted to say such 'illegal' forms as 'ten six ten' (the Korean named-value equivalent to 'sixteen tens'), but must say 'one hundred six ten', keeping to a pure named-value representation. Such sums can be found by a generalization of the over-ten method to the larger value: '8 ten plus 8 ten is one hundred (putting 2 ten from the second 8 ten with the first 8 ten) and 6 ten (left over from the original second 8 ten)'. Sums of a given value that exceed nine can also be found by using a base-ten positional representation that ignores (at least momentarily) the value of given digits and says (using an English unitary two-digit representation) '8 plus 8 is sixteen – that's one six: one to be traded and six to be recorded in this column' or (using a Korean two-digit named-value representation) '8 plus 8 is ten six: move the ten over to the next left column and record the six in this column'. Each of these Korean named-value approaches was used by some Korean second- and third-graders in explaining their multidigit addition (Fuson and Kwon, under review).

Relative difficulties in adding and subtracting multidigit numbers accurately and/or meaningfully

The lack of verbal support in the English language for multi-unit named-value/base-ten representations of tens and ones makes it particularly important that support for constructing such representations be provided in other ways to English-speaking children and other children with number words that do not clearly state the underlying tens structure of the words and the base-ten written marks. Unfortunately, in the USA, such support is rarely given. Children are taught multidigit addition and subtraction as step-by-step procedures of adding and subtracting single-digit numbers and of writing digits in certain locations. These experiences result in many US children constructing a mental concatenated single-digit representation of multidigit numbers in which multidigit numbers are viewed as composed of single-digit numbers placed next to each other (M. Kamii, 1981, called this 'glued together' digits). This representation is inadequate in many ways and results in many errors in place-value tasks and in multidigit addition and subtraction.

Children indicate use of the concatenated single-digit representation in several different place-value tasks. When shown, for example, the numeral 16 and sixteen objects and asked successively to show the objects made by each part of the numeral (the 6 and then the 1), many elementary school children indicate six objects for the 6 but indicate only one object for the 1 instead of the ten objects to which the 1 really refers (C. Kamii, 1985; M. Kamii, 1981). When asked to read a three-digit number and then write the number that is one more than the given number, half the third-graders increased by one the digit in one or more places other than the ones place: giving for 342 the answers 1342, 453, 442, 452, 352 (Labinowicz, 1985). Children also sometimes seem to use a concatenated single-digit representation to decide which of two multidigit numbers is larger: half of the third-graders sometimes ignored the position of digits and focused on a single digit in one number as being larger than a single digit in another number, choosing 198 as being larger than 231 (Labinowicz, 1985). Ginsburg's (1977) Stage 1 for children's understanding of written number – no verbalizable meaning for the digits – fits this concatenated single-digit representation: in the example protocol, the child says about the 123 just written for the words 'one hundred twenty three' that the '1 is just 1, the 2 is just 2, and the 3 is just 3'. Ross (1986, 1988) found a level of place-value knowledge in which children seem to possess quantitative meaning for the tens and ones (they relate object subgroupings of tens and of ones to two-digit numbers) but in which they actually are just relating digit values or digit positions to presented groupings without regard for the size of the group: they will say that the 6 in 26 refers to six groups of four and the 2 refers to the two left-over objects if presented with this non-ten grouping of 26 objects. Bednarz and Janvier (1982) reported for many Canadian third- and fourth-graders 'digit by digit' strategies that ignored the tens and hundreds values of the digits.

Use of the concatenated single-digit representation may not be evident when children are given multidigit addition or subtraction problems written correctly vertically (with like relative positions aligned) if the sum of the addends in each column does not exceed ten. Children add (or subtract) the digits in each column and write each sum (or difference) in the space below the column. Inadequacies in calculation performance appear if the problem is written horizontally, if the columns are aligned incorrectly, if the multidigit numbers have different numbers of digits, or if the sum in a column is ten or greater. If asked to add two multidigit numbers written horizontally, children may not even keep the digits in each given number together (Fuson and Briars, 1990) or they align the numbers on the left (Ginsburg, 1977; Labinowicz, 1985; Tougher, 1981). Friend (1979, discussed in Davis, 1984) identified several different kinds of errors Spanish-speaking children make in addition and subtraction problems in which the numbers have different numbers of digits (see Table 18.2). Further inadequacies of the concatenated single-digit representation in directing multidigit addition and subtraction are revealed when the sum in any given column exceeds nine; examples of several kinds of errors made in such problems are given in Table 18.2. For some of these errors children at least maintain the values of the single digits: if they subtract from one digit they add the same amount to some other digit. But in other errors, children do not even conserve the values of the single digits that make up the multidigit numbers.

Many US children who carry out multidigit addition and subtraction correctly do not understand multidigit numbers and do not have adequate meanings for the multidigit procedures. Many third-graders who correctly add two-digit numbers nevertheless identify the 1 written above the tens column as a 'one' and not as ten ones or as one ten (Resnick, 1983; Resnick and Omanson, 1987), and in a three-digit problem correctly added similarly identified the traded 1 as a one rather than as a hundred (for ten traded tens) despite probes such as 'What does this 1 stand for?' and 'What do the 3 and the 2 [the hundreds digits] stand for?' (Labinowicz, 1985). Only 24% of the second- and third-graders who subtracted correctly identified their trade from the hundreds place as borrowing a hundred (Cauley, 1988); the others said they had borrowed a 'one'.

Thus, there seem to be levels within the concatenated single-digit representation, both with repect to place-value performance and multidigit addition and subtraction performance. With respect to place-value performance, children may initially not even be able to label the digits as 'tens' and 'ones'. They then may begin to label digits reliably, but these digits are based on ordinal position (the name of the first column is 'ones', the name of the second column to the left is 'tens', the . . . third . . . is 'hundreds') and not on any quantitative meanings of these names. Later, some children may select grouping referents for these digits, but these grouping referents are general aspects of any given grouping rather

Table 18.2 Multidigit addition and subtraction errors that reflect a concatenated single-digit representation

Addition errors	*Subtraction errors*
Carry-to-the-leftmost[a] 11 1 6 8 1 5 6 ——— 4 1 4	Always-borrow-left[e] 2 ~~3~~ 6 15 1 0 9 ——— 1 6 6
Wrong-align-long-algorithm[b] 8 7 3 9 ——— 1 6 ——— 2 7	Borrow-unit-difference[e] 4 ~~8~~ ~~8~~ 9 1 9 ——— 3 0
Write-sum-for-each-column[c] 5 6 8 7 7 8 ——— 12 13 16	Borrow-across-zero[f] 5 ~~6~~ ~~7~~ 0 10 12 2 5 ——— 5 0 8 7
Vanish-the-one[c] 5 6 8 7 7 8 ——— 2 3 6	Stops-borrow-at-zero[e] 6 ~~7~~ 10 10 12 3 2 5 ——— 6 7 8 7
Reuse-digit-if-uneven[d] 6 3 2 ——— 8 5	Top-smaller-write-zero[g] 2 5 2 1 1 8 ——— 1 4 0
Add-extra-digit-into-column[d] 6 3 2 ——— 1 1	Smaller-from-larger[h] 2 5 2 1 1 8 ——— 1 4 6
Ignore-extra-digits[d] 6 3 2 ——— 5	Reuse-digit-in-uneven[d] 7 8 6 ——— 1 2

Note: The following are source notes: [a] Baroody (1987); [b] Ginsburg (1977); [c] Fuson and Briars (in press) and Fuson (1986); [d] Friend (1979, in Davis, 1984); [e] Van Lehn (1986); [f] Van Lehn (1986) and Davis (1984); [g] Fuson and Briars (in press) and Van Lehn (1986); [h] Davis *et al.* (1979, in Davis, 1984), Fuson and Briars (in press), Fuson (1986), Labinowicz (1985) and Van Lehn (1986).

than specific tens and hundreds groups. Within multidigit addition and subtraction performance, the concatenated single-digit representation for some children seems to serve as the basis for a spatial pattern analysis of multidigit procedures (see Van Lehn, 1986) in which the quantitative values of the single digits are not accessed, whereas for other children these values are accessed and used to constrain the trading of units from digit to digit. However, because these digits do not have tens or hundred values, these values do not constrain the columns between which trading occurs and so the errors shown in Table 18.2 occur.

Practical implications

Support for English-speaking children in constructing mental multi-unit named-value and positional base-ten representations might be provided in at least three ways. First, size embodiments such as base-ten blocks that perceptually display the relative sizes of different named values and a positional base-ten embodiment such as digit cards (cards on which a single digit is written) can be used to help children understand the features of the named-value and positional base-ten systems. Figures 18.1 and 18.2 illustrate the use of these embodiments in multidigit addition and subtraction. Some children may need to use such embodiments for a long time in order to construct the multi-unit mental representations required for multidigit numbers. During this time it is important that the named-value blocks and number words be closely linked to the positional base-ten digit cards and written marks problems so that the named-value and positional base-ten meanings can be related to each other (see Fuson, in press, for a discussion of other features of the effective use of these embodiments).

Secondly, multidigit learning/teaching might begin with four-digit numbers because the regular named-value English words for hundreds and thousands support the construction of a named-value representation and can provide a strong context into which the irregular tens can be pulled. This was done successfully for second-graders with base-ten blocks used as in Figs. 18.1 and 18.2 (Fuson, 1986; Fuson and Briars, 1990), and anecdotal evidence indicates that these larger numbers may have a similar effect for children using a multi-unit sequence representation in which they count on by tens and hundreds.

A third alternative is having children learn a 'Chinese' version of English number words, that is named-value for tens. Thus, 8653 would be said '8 thousand 6 hundred 5 ten 3' and 12 would be read as 'ten two' or 'one ten two'. These 'Chinese' number words could be introduced in a cross-cultural context and might help to focus English-speaking children on constructing a named-value rather than just a unitary representation of two-digit numbers.

Because so much multidigit school instruction occurs without sufficient perceptual or linguistic support of these kinds, it is no wonder that so much two-digit place-value and addition and subtraction instruction goes awry in the USA

SETTING UP THE PROBLEM

Thousands Hundreds Tens Ones

← 3725
+ 1647 ←

RECORDING THE ONES

Thousands Hundreds Tens Ones

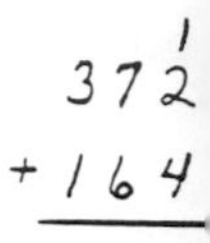
1
372
+ 164

ADDING THE ONES COLUMN

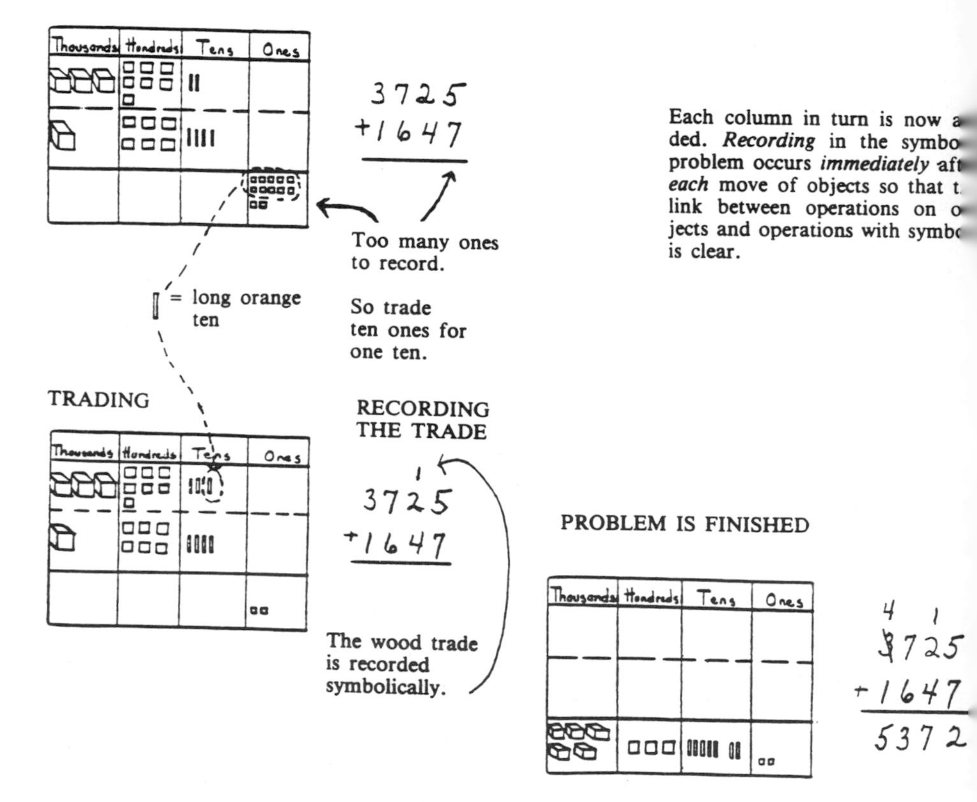
Thousands Hundreds Tens Ones
3725
+1647
Too many ones to record.
So trade ten ones for one ten.
= long orange ten
Each column in turn is now a ded. Recording in the symbo problem occurs immediately aft each move of objects so that t link between operations on o jects and operations with symbo is clear.
TRADING
RECORDING THE TRADE
1
3725
+1647
The wood trade is recorded symbolically.
PROBLEM IS FINISHED
4 1
3725
+1647
5372

Fig. 18.1. Multidigit adding and recording with base-ten word embodiment.

ETTING UP THE PROBLEM

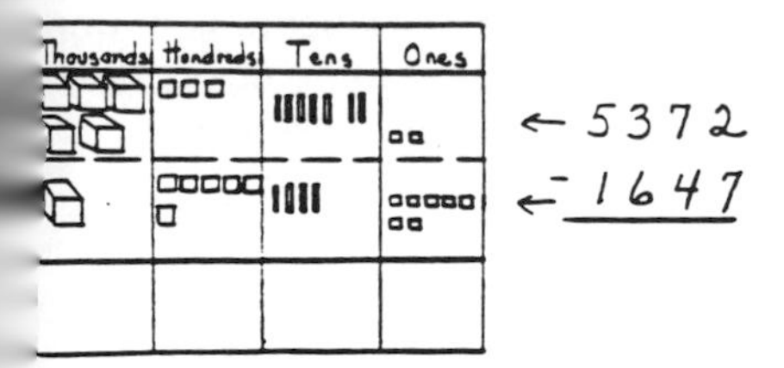

FIRST: Do all trading

Need more ones in order to subtract. Trade one ten for ten ones.

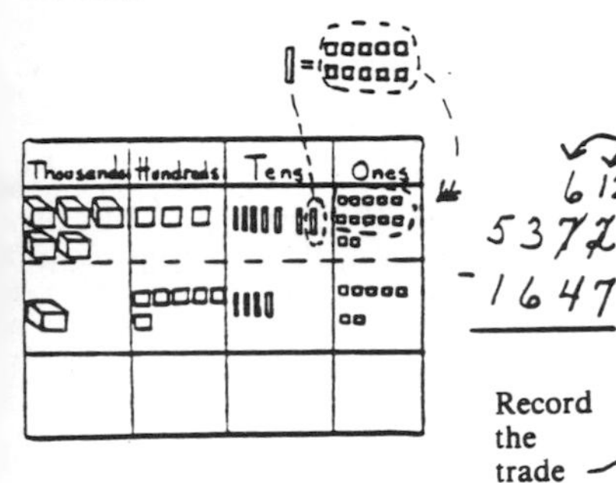

Now make each necessary trade, recording *each trade* on the symbolic problem *immediately after* trade is made.

SECOND: Do all subtracting

RECORD COLUMN BY COLUMN

SUBTRACT THE ONES

RECORD

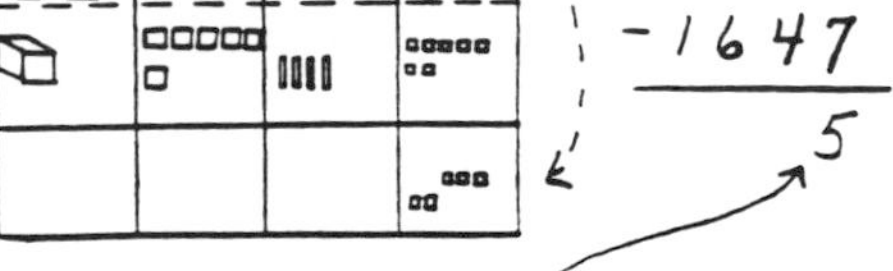

To subtract 12,7, a child can
count up from 7 to 12,
know the fact,
go over 10 (7 + 3 + 2),
or any other method.

Each column in turn is now subtracted. *Recording* in the symbolic problem occurs *immediately after each* move of objects. Each column is recorded before the next column is subtracted with the wood.

PROBLEM IS FINISHED

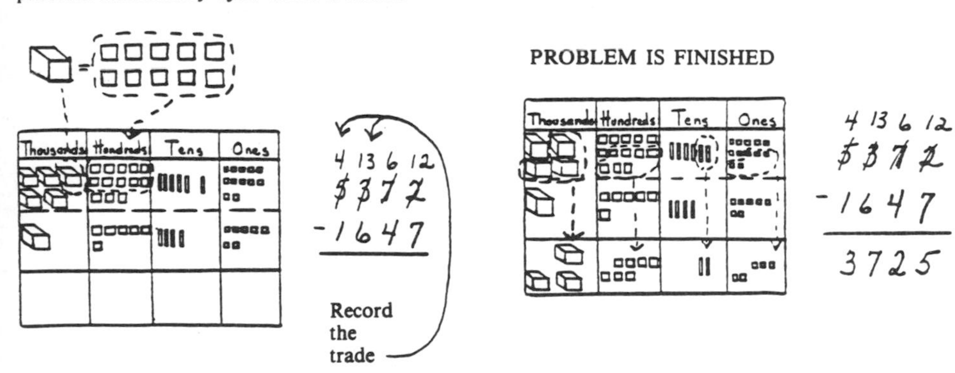

Fig. 18.2. Multidigit subtracting and recording with base-ten word embodiment.

(and possibly in many European countries): teachers talk about tens and ones but children see the written two-digit marks as unitary sequence/counting numbers that are counted collections of single objects (or words), or see these marks as concatenated single digits, each with only a unitary meaning.

Conclusions

Asian languages that have a regular named-value system of number words that name the ten values in a regular way as well as naming the hundred and thousand values, help Asian children construct multi-unit mental representations for multidigit numbers. These mental representations allow Asian children to add and subtract numbers with sums between 10 and 18 and to add and subtract multidigit numbers earlier, more easily and more accurately and render multidigit addition and subtraction more meaningful. English-speaking children construct and use for a long time unitary representations of number instead of multi-unit representations and are much more likely to construct an inadequate concatenated single-digit representation of multidigit numbers that allow them to make errors in place-value tasks and in multidigit addition and subtraction. Because of the lack of support for understanding tens and ones in English, perceptual or linguistic support for constructing adequate multi-unit representations needs to be provided in the classroom.

Further reading

Fuson, K.C. (submitted). *Children's representations of multidigit numbers: Implications for addition and subtraction and place-value and teaching.*

Labinowicz, E. (1985). *Learning from children: New beginnings for teaching numerical thinking.* Menlo Park, Calif.: Addison-Wesley.

Menninger, K. (1969). *Number words and number symbols: A cultural history of numbers* (translated by P. Broneer). Cambridge, Mass.: MIT Press. (Translated from original publication, *Zahlwort und ziffer*, 1958, Vandenhoeck & Ruprecht.)

19 Learning to count in Japanese

Hajime Yoshida and Kazuhiro Kuriyama

Although counting was regarded for a long time as a mere rote skill, recent research has increasingly demonstrated its importance as a basic necessity in the understanding of numbers by young children (see Chapters by Fuson and by Cowan, this volume, for a fuller discussion). For this reason, we need to know more about how children progress through the early stages of learning to count, and collect data on children learning numbers in different languages.

This chapter is concerned with the ways in which children learn to count, and exploit counting, in the early stages of numerical development. Counting is a verbal skill which is closely related to how young children represent numbers. In the main part of this chapter, we present summaries of a series of our own studies, conducted with Japanese children, that are concerned to uncover the relative status of the different number words for the early counter. We propose that the numbers 1–5 serve as a firm base in young children's number systems. We discuss this with reference to the beliefs concerning number acquisition that influence Japanese educational practices, though it will be clear that our arguments may hold for children learning to count in other languages as well. We consider briefly some of the differences in the ways in which numbers are organized in Japanese compared to English, and draw on findings which indicate that these linguistic differences have direct consequences for the learner. Finally, we suggest some practical implications.

Almost all maths textbooks in Japan have for a long time regarded counting as an activity which interrupts a conceptual number understanding with an emphasis on its mechanical aspects. Although counting itself, of course, has such aspects, new and interesting aspects of counting have recently been discovered. For example, the concept of conservation has been shown to be related to accurate counting (Fuson *et al.*, 1983). In addition, early competence for computational algorithms, such as addition and subtraction, has been shown to depend upon counting skills (Woods *et al.*, 1975). Such studies suggest that counting is not merely a mechanical activity, but reflects and perhaps promotes numerical knowledge in children.

Counting is mainly used in two ways by young children. First, counting is employed to count concrete objects. Some basic aspects of counting are manifest in this kind of counting. One basic aspect is that counting to assess the cardinal numbers of a set is governed by five principles (Gelman and Gallistel, 1978): one–one correspondence, the stable order principle, the cardinal principle, the item-indifference principle, and the order-indifference principle.

The other way in which counting is used is through counting with one's fingers or orally without the need for concrete objects. Previous research has not elucidated the principles underlying this mode of counting. However, we have found some evidence of the principles behind such counting. Our studies suggest that counting orally or by using one's fingers also reflects the representational structure of smaller numbers. In particular, we confirmed through three studies that preschool children represent the numbers 1–5 as a privileged anchor or firm structure for numbers below 10. Previous research had investigated the basic aspects of counting. In addition to these aspects, our work examined the representation of number concepts derived from counting.

Our first study investigated the structure of the number concept by analysing the strategies employed in solving both addition and subtraction problems (Yoshida and Kuriyama, 1986; Kuriyama and Yoshida, 1987). The second study confirmed such a representational structure as a privileged anchor by analysing forward and backward counting (Kuriyama and Yoshida, 1988). The last study examined further such representational structures through tasks in which some numbers are resolved into x and y (Yoshida and Kuriyama, in prep.) Let us now summarize these studies.

The numbers 1–5 as a privileged anchor

In our first study, we aimed to find evidence for strategies which might have been related to the representation of the numbers 1–5 as a privileged anchor in children's processes for solving both addition and subtraction problems. Further, if children have such a representation of the numbers 1–5, we might expect evidence of such a representation in incorrect strategies when children are solving both types of problem.

The subjects included 29 children (14 boys and 15 girls) attending a private kindergarten in Miyazaki and they came primarily from middle-class homes. We set the children a number of addition and subtraction problems. The sums of the addition problems were 10 or less. The problems were presented in such a way that the larger number was always the augend. The minuend and subtrahend of the subtraction problems were always 9 or less. Video cameras and video-tape recorders were used to record the activities of the children when solving the addition and subtraction problems. An experimenter told the children that they would be allowed to use their fingers when responding to the questions. During the practice phase, if any child made a mistake, the experimenter gave them feedback. In the test phase, the experimenter gave no feedback to the child even if he or she answered incorrectly.

Four types of strategies for solving both addition and subtraction problems were found to be used, either from the children's verbal responses or from the video-tape:

1 *Overt counting*: the children counted numbers by opening their fingers overtly one by one (O-type).
2 *Covert counting*: the children counted by opening their fingers, by following their fingers with their eyes, by moving their heads covertly, or by moving their head slightly (C-type).
3 *Direct representation of numbers*: the children represented numbers on their fingers directly without counting their fingers one by one or covertly (D-type).
4 The last was a type in which we were not able to observe any visible strategy. When questioned, the children who had used this type responded that they had done the problems in their heads. Therefore, this type was called *internal representation* (I-type).

We obtained some evidence about the importance of the numbers (1–5 as a unit or structure by analysing these strategies, especially O-type counting. When dealing with single-digit numbers above 5, the children counted overtly the numbers in one of two ways: first, they counted their fingers one by one (O-all); secondly, they opened their fingers directly to 5 without any counting, then one by one for the remaining numbers over 5 (D-to-O). Table 19.1 shows the mean numbers of problems in which children used the different strategies for addition and subtraction. As Table 19.1 shows, the D-to-O type was used more frequently than the O-all type for addition.

Furthermore, we found that children adopted different strategies for solving problems with numbers less than 5 compared with those used for solving numbers greater than 5. By dividing into one or other of these two problem categories, we analysed strategies for all the addition and subtraction problems as combinations of these four counting types. We found 12 combinations of strategies for the addition problems and 14 for the subtraction problems. The first letter of each strategy indicates the strategy which was used for the first addend and the second letter the strategy which was used for the second addend.

Table 19.1 Mean numbers of problems for two overt counting types in addition and subtraction

	Overt counting type	
Problem	*O-all*	*D-to-O*
Addition	1.86 (1.29)	3.00 (1.06)
Subtraction	1.44 (1.34)	10.67 (7.87)

Note: See text for explanation of O-all and D-to-O. Standard deviations are in parentheses.

Table 19.2 Frequency ratios for problems with numbers below and over 5 in addition and subtraction problems

	Addition problems		*Subtraction problems*	
Strategy	*Below 5*	*Over 5*	*Below 5*	*Over 5*
D-D	0.548 (41.33)	0.328 (33.52)	0.496 (32.57)	0.284 (29.01)
I-I	0.230 (32.93)	0.201 (33.52)	0.140 (26.00)	0.152 (22.14)
D-I	0.103 (22.55)	0.072 (12.85)	0.149 (28.50)	0.105 (25.23)
C-C	0.071 (17.48)	0.096 (21.90)	0.090 (15.92)	0.086 (13.96)
O-D	0.000 (0.00)	0.100 (20.44)	0.105 (17.08)	0.083 (12.61)
D-O	0.013 (4.70)	0.057 (11.74)	0.000 (0.00)	0.092 (19.19)
O-O	0.000 (0.00)	0.045 (11.17)	0.016 (5.59)	0.099 (18.25)
C-I	0.019 (7.51)	0.027 (5.37)	0.020 (6.38)	0.054 (15.52)
D-C	0.010 (3.35)	0.027 (5.37)	0.020 (9.24)	0.038 (10.30)
O-C	0.003 (1.47)	0.036 (11.88)	0.000 (0.00)	0.012 (2.91)

Note: Standard deviations are in parentheses.

For example, a strategy of D-D in the problem 6+2 means that children first opened their six fingers directly without overt or covert counting, and then added two fingers directly to the six fingers before responding 8. The frequency ratios of the seven most common combinations to the total number of problems in each category are shown in Table 19.2.

The ratio of D-D for addition problems increased significantly for problems with numbers below 5 compared to those with numbers over 5. There were, on the other hand, significant decreases in the ratios of O-D and D-O for addition problems with numbers below 5 compared to those with numbers over 5. For subtraction problems, the ratio of D-D increased significantly for problems with numbers below 5 compared to those with numbers over 5. However, there were significant decreases in the ratios of O-D and O-O for problems with numbers below 5 compared to those with numbers over 5.

We postulate that the counting strategies developed from the overt to the internal through direct or covert strategies. Based on this assumption, the above results would indicate that children relied on developed strategies for problems with numbers below 5. However, they utilized more primitive strategies when solving difficult problems.

Children's errors in these tasks were mainly of two types: those that contained errors related to the number 5, and those that contained errors unrelated to the number 5. These types are shown in Table 19.3. An interesting error was observed for the numbers $\geqslant 6$ (1a). When children represented such numbers with five fingers in one hand and x's in the other hand, they then closed the hand with the x's, forgot them, opened their hand again with the number of fingers to be added, and added that number to 5. This kind of error could be explained in terms of the representation of numbers to 5 as a privileged anchor. That is,

Table 19.3 Types of errors found in study 1

Category	*Frequency ratios*
1 Errors related to the number 5	
(a) In representing the number 5, children have 5 in one hand and the remaining number (x) in the other hand. They then forget x	0.16
(b) In representing the number 4, they have four fingers in one hand. The remaining fingers are then opened, and they think the number is really 5	0.04
2 Errors unrelated to the number 5	
(a) Answers of problems were +1 from the correct answers	0.40
(b) Addend or augend, minuend or subtrahend were used as answer	0.10
(c) In subtraction problems with numbers below 5, children represented the number in one hand. They show the subtrahend in the other hand but merely close this hand, leaving the other fingers as the answers	0.07
(d) Problems solved using the task opposite to the correct one	0.05
(e) Not clear	0.18
(f) Could not answer	0.04

because numbers over 5 may be a rather obscure structure for young children, they may have forgotten the numbers over 5.

External *vs* internal representation

The results of our first study suggested the possibility that children represented the numbers 1–5 as a privileged anchor. However, because there are five fingers on each hand, there may be another possibility, i.e. such a representational system comes from the external structure of the fingers and does not reflect the internal knowledge structure of children. Therefore, a second study was designed to investigate whether or not the results obtained with the first study could be due to internal representation.

To do so, we needed to adopt some task in which children do not need to depend upon their fingers. A counting task was considered to be the best to investigate such representation. An adult is able to start counting at any number and to stop it at any number. Very young children might not stop counting at a given number, but continue to count until they reach the final number which they know. Therefore, if they could count from 1 to a given number 'b' correctly, it would mean that they understood that number-sequence and could divide into numbers from 1 to 'b' and into numbers over 'b'. This may be the third step in the development of counting skills (Fuson *et al.*, 1982).

If children represent the numbers 1–5 as the privileged anchor, they should easily stop their counting at these numbers in both forward and backward counting. Further, it should be easier for children to stop at any number below 5 than when asked to stop at a number over 5.

We tested these possibilities with 4- to 5-year-old children (13 boys and 12 girls) attending a private kindergarten in Miyazaki City. The children were given 14 forward counting and 16 backward counting problems. The numbers 1–9 were used. The experimenter told the children that today's game was playing with numbers. She told them not to use their fingers to find the answers to the questions. First, each child was asked to count from 1 to 9. Feedback was given to incorrect responses in this practice phase. However, in the test phase, no feedback was given. With forward counting, the following tasks were given: to count from 2 to 4 (abbreviated as 2–4), 2–5, 2–8, 2–9, 3–5, 3–7, 3–9, 4–7, 4–9, 5–7, 5–9, 6–8, 6–9, 7–9. The backward counting tasks were: 3–1, 4–1, 4–2, 5–1, 5–2, 5–3, 6–3, 6–4, 7–3, 7–5, 8–3, 8–5, 8–6, 9–3, 9–5, 9–7.

If the numbers 1–5 may be represented as a firm structure, then we would expect a difference between the performance on the problems with numbers below 5 and the problems with numbers over 5. All the forward and backward counting problems were divided into two categories: those problems that contained numbers $\leqslant 5$ and those problems that contained numbers > 5. The mean ratio of correct responses for the two types of problems with forward counting were 0.86 and 0.79, respectively, showing no significant difference. The mean ratio of correct responses for the two types of problems with backward counting were 0.75 and 0.35, respectively, a statistically significant difference. Although there was no difference between the two types of problems with forward counting, the > 5 problems were clearly more difficult than the $\leqslant 5$ problems with backward counting.

To investigate further, we examined the children's errors. The counting errors were divided into seven types:

1 Those where the children only counted the first and last number.
2 Those where the children omitted one or some number words when counting.
3 Those where the children repeated the same number word.
4 Those where the children always began counting from 1.
5 Those where the children did not stop at the last number (cannot-stop type).
6 Those where the children counted forwards in backward counting tasks.
7 Random counting (random type).

Of these error types, two stand out. The first is where the children were unable to stop counting at the requisite number. All of the children were able to stop at the stopping point '5' when counting forwards, but many children were unable to stop counting when the stopping point was a number other than 5, i.e. many of the children were unable to split the number-word sequence into two parts at any number other than 5. The other interesting category was the random errors. These errors were not found in responses to the $\leqslant 5$ problems, but they were found with > 5 problems. These results indicate that the numbers 1–5 function as a privileged anchor. Also, the results suggest that the role of the numbers 1–5 is not restricted to the external structure of the fingers, but also to the internal representation system.

The nature of the anchor: 1–5, or 5 alone?

The two studies thus far confirm that preschool children use the numbers 1–5 as a privileged anchor in the numbers below 10. However, the results obtained from the second study raise another possibility, i.e. whether or not the anchor is the numbers 1–5 or just the number 5 alone. In the second study, all of the children were able to stop counting at the number 5 in forward counting tasks. If the children represent all of the numbers 1–5 as the privileged anchor, they should also have stopped when required at the numbers 2, 3 and 4, but they did not do so.

These results suggest that the privileged anchor consists of the number 5 alone. Therefore, in a third study, we tested for this by resolving some numbers into x and y. If the number 5 alone is represented as the privileged anchor, children should be able to resolve numbers into 5 and x more easily than numbers other than 5 and x. In addition, if children have such a representation, we should find some differences in the strategies used when the task requires them to resolve by 5 or by a number other than 5.

In this study, we tested 20 five-year-old children (10 boys and 10 girls), all of whom attended a private kindergarten in Miyazaki. All of them were allowed a practice session before testing on each resolve task was carried out. For example, in the practice phase, the experimenter said '9' and asked the children to resolve it into 2 and x. When all of the children understood how to resolve the problem, the test phase began. Each child was randomly allotted the numbers 3, 4, 5, 6, 7 and 8, and asked to resolve them into 2s and xs (abbreviated as resolve 2 task). Then, the children were given their next resolve task. There were five problems (the numbers used were 4, 5, 6, 7 and 8) in the resolve 3 task, four problems (5, 6, 7 and 8) in the resolve 4 task, three problems (6, 7 and 8) in the resolve 5 task, two problems (7 and 8) in the resolve 6 task, and one problem (8) in the resolve 7 task. The order of presentation of the resolve tasks was randomized.

The mean ratio of correct responses for the resolve 2, 3, 4, 5, 6 and 7 tasks were 0.69, 0.80, 0.76, 0.93, 0.75 and 0.65, respectively. These differences demonstrated that preschool children use 5 alone as the privileged anchor for the numbers below 10.

An analysis of the children's strategies revealed that some were more prevalent than others. In particular, the main strategy used for the number 5 was the D-type direct representation of numbers, whereas the main strategy used for numbers other than 5 was the O-type overt method of counting.

If children represent all of the numbers 1–5 as the privileged anchor, they should operate on such numbers by using a direct strategy and not an overt one. We believe that children are easily able to present these numbers as the privileged anchor by using their fingers, because such numbers are organized very clearly.

We do not believe, however, that children operate on such numbers by counting their fingers one by one if the numbers function as an anchor.

Japanese education and the privileged anchor

Some readers may argue that our results on the role of the number 5 may be unique to Japanese culture. In Japan, we use an abacus (a *soroban*) which has a 5-unit counter and four 1-unit counters for each place value in the decimal system (see Fig. 19.1). When operating the *soroban*, one is able to achieve a result using the counters, and need not internally store the results. Therefore, skilled users are able to operate with large numbers with accuracy and can do without the abacus, relying on their mental abacus instead.

Our work suggests that children represent the number 5 as a firm structure or a privileged anchor. Our findings are congruent with the fact that the number 5 is used as an important basic unit with the abacus. However, we believe that children's representation in our work does not reflect an effect of the abacus. In general, children start to learn the abacus in the third grade of Japanese elementary schools. However, our subjects were all preschool children. Although they may have seen an abacus in their homes, they would have been prevented by their parents from using it before entering elementary school.

The Japanese language and further developments in counting

As Fuson and Kwon (this volume) have already discussed, there are some differences between Eastern and Western languages in terms of learning to count. We echo their arguments with reference to Japanese. First, Japanese number words are more regular and systematic than their Western counterparts. The pronunciation system in Japanese number words is congruent with the base decimal number system. Such systems are shown in Table 19.4. In English number words, 11 is pronounced as eleven, 12 as twelve and 13 as thirteen. However, in Japanese number words, 11 is pronounced as jyu-ichi (in English this is the same as ten-one), 12 as jyu-ni (ten-two) and 13 as jyu-san (ten-three). In this system, Japanese number words have the elements of tens and ones. However, English number words lack such elements. Therefore, as Fuson and Kwon have elaborated, a system such as Japanese might assist children in developing cognitive structures of number that reflect the decimal number system.

Secondly, the number words in Japanese are pronounced in the same order as their written form. Japanese number words are always written with the tens first and then the units, and they are pronounced in the same order. For example, as shown in Table 19.4, 17 is pronounced as jyu-shichi (ten-seven), 18 as jyu-hachi (ten-eight) and 19 as jyu-ku (ten-nine). In the English language, however, this is not the case: 17 is pronounced as seventeen, 18 as eighteen and 19 as nineteen.

Table 19.4 The pronunciation system in Japanese number words

1	ichi	11	jyu-ichi	21	ni-jyu-ichi
2	ni	12	jyu-ni	22	ni-jyu-san
3	san	13	jyu-san	23	ni-jyu-san
4	shi	14	jyu-shi	24	ni-jyu-shi
5	gou	15	jyu-gou	25	ni-jyu-gou
6	roku	16	jyu-roku		
7	shichi	17	jyu-shichi		
8	hachi	18	jyu-hachi		
9	qyu	19	jyu-qyu		
10	jyu	20	ni-jyu		

Although the number words are also written with the tens first and then the units, they are pronounced with the units first and then the tens. Thus, number words in Japanese correspond exactly to their written form.

However, for numbers over 20, there is another difference between Japanese and English in the order in which number words are pronounced and written. In the Japanese language, number words over 20 are pronounced in the same order as the construction of the numbers. For example, 24 is pronounced as ni-jyu-shi in Japanese (in English this should be pronounced as two-ten-four) and 32 as san-jyu-ni (equivalent to three-ten-two in English). Japanese number words have a clear number system even with number words over 20. Numbers are mentally represented and stored through language. Therefore, Japanese children have less difficulty in expressing printed numbers orally and are less confused when processing them mentally, compared to children whose mother tongue is English.

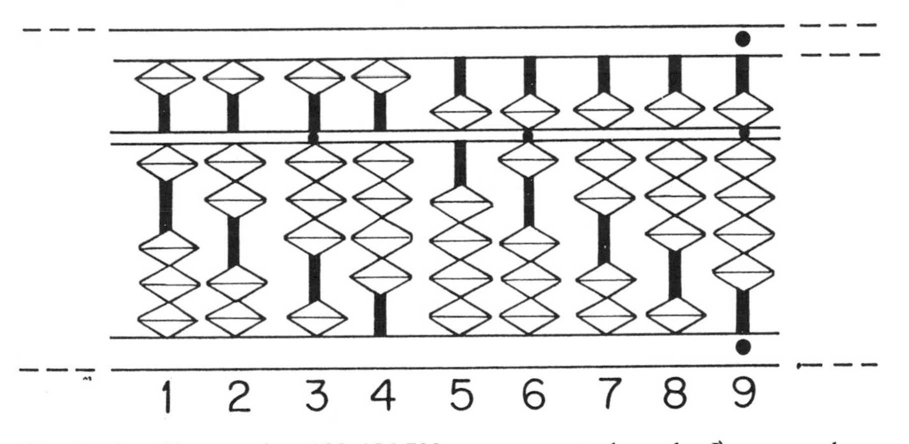

Fig. 19.1. The number 123,456,789 as represented on the Japanese abacus.

Practical implications

Our work shows that there is an intermediate period before children acquire the decimal number system. In terms of acquiring this system, we would postulate three phases of development: (1) gaining perceptual understanding, (2) using the number 5 as the privileged anchor and (3) the development of the decimal number system.

How can such a developmental sequence be applied to everyday instruction? At present, almost all instruction in number concepts is centred on the decimal number system. For example, children learn this system through composition or resolution tasks at the beginning of elementary school. These tasks are very difficult for the first-grader to comprehend. Therefore, we recommend that these tasks should be taught using the number system based on the number 5, i.e. 7 is divided into 5 and 2, or 8 into 5 and 3. If children master these kinds of tasks by using the number 5 as an anchor, they will transfer easily from the system based on the number 5 to the decimal number system because the former provides stable knowledge that children will have acquired through everyday life.

In Japanese elementary schools, the number system based on the number 5 has been used in a unique way, called *suidou-houshiki* ('water-supply method' in English). With this method, the numbers are represented by tiles which are cut out of cardboard. When numbers greater than 5 but less than 10 are being demonstrated, the tile which represents the 5-unit is all important: 6 is represented as one 5-unit and one-unit, 7 as one 5-unit and two 1-units, etc. The use of the 5-unit tile is helpful when children understand the number concept.

Furthermore, we think it can be advantageous for children to solve elementary addition and subtraction problems by using the number 5. For example, in the problem $7 + 8$, the child should resolve the number 7 into 5 and 2, and 8 into 5 and 3, and then add the 5 of the 7 and the 5 of the 8 ($= 10$). Next, she or he should add the remaining numbers together, i.e. the 2 of the 7 and the 3 of the 8 ($= 5$), and then the 10 and the 5 which have already been computed. This way of working could be helpful to children for accurate calculation.

Conclusions

Counting is one of the basic cognitive skills in the development of number concepts in children. By analysing this skill in children, we found that there is a firm representational structure based on the number 5 in the numbers below 10. Such findings are the results of a cognitive psychological approach. However, we believe that such results have direct implications for instructional intervention. The number system based on the number 5 would be helpful for the maths curriculum in kindergartens as well as for the early grades of elementary schools. In fact, one of the typical maths curricula (water-supply method) in Japan takes wholly into account children's representational structure based on the number 5. This kind of collaboration between research and practice has a considerable role to play in the improvement of practical instruction.

20 Mathematics in a multilingual society

Philip C. Clarkson

Introduction

Atawe began his first year of schooling at about the age of 8. He was taught by a Papua New Guinean who usually spoke Melanesian Pidgin, and sometimes English. Both languages were unknown to Atawe. A local teenage boy who spoke some 'Pidgin' translated for the teacher into Atawe's own local language. The following year Atawe graduated to the next grade. This grade was taught by an American missionary who taught in English with a rare utterance in 'Pidgin'. For many days Atawe sat in a fog wondering what was being said.

Atawe's early school experiences highlight the language difficulties all students in Papua New Guinea (PNG) have when they attend school. In a country of $3\frac{1}{2}$ million people, where 16% of the world's languages are spoken, it is not surprising to find that the effect of language on learning has been a research topic for a number of years. English is the official language of communication in schools. 'Melanesian Pidgin', and to a lesser extent 'Police Motu', are common lingua franca. In the villages where the vast majority of people live, the local or village language (*Tok Ples*) is spoken. More than 720 different *Tok Ples* have been noted; that is 720 distinct languages, not dialects. Each language typifies a different culture.

In this chapter, consideration will be given to the specific difficulties faced by pupils learning mathematics in a multilingual society such as PNG. After summarizing the background political considerations influencing the directions of education in this multicultural society, a brief review of recent research which points to the importance of linguistic issues in PNG education will be given. Following this, an account of a series of studies designed to investigate PNG students' difficulties with mathematical word problems will be presented. The chapter is drawn to a close with a discussion of some practical implications of the studies and suggestions for future research.

Education and national goals in Papua New Guinea

The politicians have decided that for PNG to become an integrated nation it must develop its own modern technological culture, akin to that of Western cultures, but based on its own ancient multicultural traditions. The education system is seen as a fundamental tool in producing this new culture. Sixty per cent of PNG children now attend community school (years 1–6), the primary level of the system. At the end of grade 6, national examinations help determine which children will progress to the second level, provincial high school (years 7–10). Overall, there is a 60% transfer rate from year 6 to 7. At the end of year 8, 40% of children are requested to leave school on the basis of school-based assessment procedures. At the end of year 10, a further set of national examinations is held. Depending on how successful students arc, they can progress to a variety of tertiary level institutions with vocational orientations, national high school (years 11–12), or a preliminary year in some of the departments at one of the two universities. Attendance at a national high school is becoming the normal route to the universities. National examinations are held again at the end of year 12.

It is clear that PNG students must overcome various examination hurdles if they are to continue through the education system. At each hurdle, mathematics is a compulsory subject. It is quite clear that mathematics is regarded by the nation's planners as being fundamental to the country's progress. It has also become clear in the last 10 years that the interaction between language learning and mathematical learning in PNG is an area of great concern.

The role of English and mathematical performance

A number of general studies have clearly indicated that a student's command of English plays a role in his/her performance in mathematics. Souviney (1983) tested students in grades 2, 4 and 6 with various language and mathematical instruments, and on eight measures of cognitive development. His results showed that English reading and Piagetian measures of conservation were highly correlated with mathematical achievement. Correlations between memory measures and mathematical achievement decreased from grades 2 to 6, while correlations between measures of language and cognitive development, and mathematical achievement tended to increase. It was hypothesized that as the mathematical tasks became more complex and hence the symbolic load increased, reliance on memory skills, in particular visual ones, was not adequate. The more successful students in higher grades were able to utilize their language abilities.

Clements and Lean (1981) investigated various cultural, memory, spatial and mathematical variables in a cross-cultural study. Students were drawn from four different community schools and an international primary school in Lae. The authors found no educationally significant differences between the PNG students and the mother-tongue English-speaking students on memory tasks, some of the mathematical tasks and on Piagetian conservation tasks. On some spatial tasks, on pencil-and-paper tests of mathematics and on a mathematical language test, there were differences. In particular, the authors noted that the PNG students handled word-free computational problems quite well, but with verbal arithmetic problems they had great difficulty. This they thought was because of the students' general difficulty with English, and especially with mathematical English.

Other studies have started to map in the detail of the interrelatedness of English and mathematics achievement. Meek and Feril (1978) tested 171 grade 6 students on 11 mathematical words drawn from the grade 5 syllabus, and presented in three differing contexts. The students had a test average of 44%, and for only one of the words did more than a third of the students obtain a correct answer in all three contexts. Jones (1982) investigated the acquisition of the concept more–less by students in grades 2–10. His results indicated that although both PNG students and mother-tongue English-speaking students learned the three contextual meanings of this concept in the same order, the PNG students were up to 4 years slower. Twenty five per cent of PNG students still had not mastered the indirect sub-construct by year 10. It was found that virtually all English speakers had done so by year 6. The implications of this are worrying in that the PNG mathematics syllabus assumes mastery of more–less by level 4, as well as the subsequent ideas of bigger–smaller, shorter–longer, in all their complex forms.

The three studies of the author's outlined in this chapter further confirm the link between language learning and mathematics learning, and add extra detail. They all concentrate on the solving of mathematical word problems, a widely recognized area of difficulty for many pupils (see De Corte and Verschaffel, this volume).

Types of pupils' errors

The central instrument of the first of the three studies used a model which delineates a hierarchy of steps through which it is proposed a student progresses in solving mathematical word problems. The steps are set out in Table 20.1, with accompanying statements which an interviewer would use when investigating at which stage an error may have occurred. The original model was devised by Newman (1977) and has been used in a number of studies (Clarkson, 1980, 1986; Clements, 1980, 1982; Newman, 1981; Watson, 1980). This model enables a specific classification of errors to be made. Three of the categories – reading, comprehension and encoding – are overtly language-based, the first two being steps in the decoding process of the written problem, with the fifth category being the reverse process.

A battery of tests including a general mathematics test containing items on logic, spatial ability and numeracy, a mathematical language test and a sheet on which students indicated their confidence of obtaining correct answers was

Table 20.1 Error categories and the beginning interviewer's statements derived from Newman (1977)

Reading	Read the question to me.
Comprehension	What is the question asking you to do/find?
Transformation	How will you do the question?
Process skills	Complete the question for me.
Encoding	Write down the answer.
Careless	

Table 20.2 Frequency of error types for 95 grade 6 children (after Clarkson, 1983)

Error category	*% Frequency*
Reading	12
Comprehension	21
Transformation	23
Process skills	31
Encoding	1
Careless	12

administered to 95 year 6 students in two urban community schools. The procedure used to categorize the types of errors the students made on the general mathematics test was similar to that followed in other studies that made use of the model proposed by Newman. More details of the procedure and instrumentation can be found in Clarkson (1983).

The frequency of error types are shown in Table 20.2. A third of the errors made are in the first two categories. Examples of reading errors were 'spaces' for 'shapes', 'meaning' for 'missing' and 'volume' for 'value'. Some of the key terms which gave rise to specific comprehension errors were 'value', 'simplest form', 'and', 'on', 'not', 'more than' and 'less than'. Comparison of the data in Table 20.2 with Australian data suggest that PNG students made more reading/comprehension errors than their Western counterparts.

In the recent past in PNG, as elsewhere, there have been calls for students to be drilled in basic number operation (process skills) with the clear impression created that this is all that is required to improve mathematical standards (e.g. Edwards and Bajpai, 1979). These results clearly show that the situation is far more complex than that simple solution implies.

An examination of the correlations between test scorers indicated that students who were taking a relatively long time to complete the general mathematics test were having difficulty both in reading and comprehending the items. Also, students who made fewer reading errors had higher scores on the language mathematics test and often predicted the correctness of their answers on the general mathematics test. This pattern did not occur when the comprehension errors were considered and hence suggested that reading and comprehension are two distinct abilities which must be mastered. However, only initial errors were noted in this study when the hierarchy was being worked through during the interview. It may be that a different pattern would emerge if all errors were recorded.

Word problems in grade 5

Following on from the study described above, a bank of 96 word problems was composed for use in further research projects. The bank was developed in collaboration with PNG research officers with local teaching experience and administered to 128 year 5 students along with a number test. The set of word questions was also distributed to community school teachers with a questionnaire asking them to indicate any questions they would not use in their classrooms, and whether there were any questions which were too difficult.

An examination of the teachers' responses showed that they considered all questions suitable for their classrooms, although three questions were thought to be rather difficult. Interestingly, the teachers did not refer to language as a source of this difficulty.

An analysis of the students' results also proved instructive. It was clear that they had found both the word and the number tests quite difficult, more so than the word test. This was quite at variance with what the teachers had predicted. There was quite a high correlation (average 0.69) between the corresponding items on each test, which was to be expected (see Clarkson, 1984a). Examining the item difficulties, on average two-thirds of the students obtained a wrong answer on each word item and half the students were wrong on each number type problem. Interestingly, for 15% of the items, the number question had a higher item difficulty than the corresponding word item. No common feature such as the operation type seemed to explain the observation. It does indicate, however, that some care needs to be taken with the general statement of 'word items are harder'.

In referring to arithmetic operations, there is a variety of words that can be used. For addition, there is 'add', 'added to', 'plus', 'sum of', and a number of others. For students to work with ease with any operation they should be familiar with the range of different terms that could be used. The test items used a number of different terms for each operation. When the students' results were examined, there was an indication that they were only familiar with one or two terms, which perhaps their teachers had constantly used. It would have been hoped that by the end of grade 5, the students would have been in command of a range of vocabulary dealing with the addition and subtraction operations, and possibly the multiplication operation.

A further analysis of pupils' errors

The results of the first study seemed to give an indication that the model devised by Newman may be a useful tool to investigate PNG students' mathematical learning. It was decided to use that model again in a further study. Twenty questions from the 96 trialled in the second study were selected. A total of 302 grade 6 students from five schools were tested and 1521 errors were investigated. Table 20.3 shows the resulting error profile of these students.

There is no space in this chapter to discuss at length the transformation, process skills and careless errors that were made by the students. Suffice it to say that with regard to the process skills, at which most of the trivial 'back to the basics' calls are directed, these data suggest that students who did not make an error in getting to that point normally solved the item.

With the encoding category, the reading and comprehension categories can be regarded as being language-based. There was an increase of 6% in the combined number of reading and comprehension errors when compared to the first study. Taken separately, however, there was a moderate fall in reading errors, but a large increase in comprehension errors. With the change in the type of items between the two tests, the second using far more words than the first, it was expected that the frequency of errors in both categories would have increased. The decrease in the reading category possibly came about because of the fewer special terms used. There were a number of items which could be called 'story problems' which perhaps helped the reading. Also, all the items had been written in PNG and this perhaps prevented the use of unusual word constructions for these students.

It is instructive to look at some examples of the specific symbols which caused comprehension errors. One of these was the decimal point. Most students referred to it as 'decimal' or 'point'. Some called it a 'full stop'. In money questions, practically without exception, students explained its use by saying it separated the toea from the kina [100 toea (t) equals one kina (K)]. In such a context, this is probably all right. However, when pressed for a broader definition, or asked for an explanation in a non-money context, many students could not give one. Interestingly, on another 'number test', when students were asked to add 2.16 and 3.70, a 'K' would often appear in front of the answer. Perhaps the context of money was so powerful students could not go beyond it. Other evidence that students did not have a broad concept of what the '.' stood for surfaced when students obtained an answer with more than three digits. Decimal points would often start appearing.

It was also found that the symbols for fractions caused difficulties. One of the items on the test was 'Kapi went shopping with K9.00. He spent $\frac{1}{3}$ of his money on food. How much did he spend on food?' Even if students could not manipulate the algorithm '$\frac{1}{3}$ of 9', it would have been hoped that towards the end of grade 6 they could have obtained the correct answer with the idea of sharing '9' into '3 equal parts'. Some students used this approach. However, many students when asked what $\frac{1}{3}$ was, or how do you get $\frac{1}{3}$ of something, had no ready answer. In

Table 20.3 Frequency of error types of 302 grade 6 students

Error category	*% Frequency*
Reading	5
Comprehension	34
Transformation	24
Process skills	14
Encoding	7
Careless	16

further probing, a rectangle was drawn and students were asked to imagine it as a piece of wood. They were then asked how we could get $\frac{1}{3}$ of it, and where we would cut it. A few suggested cutting it into three pieces, but invariably the pieces were clearly of unequal area. In this respect, PNG students manifest difficulties with the representation of fractions akin to those experiences of children elsewhere, as discussed in more detail by Kerslake (this volume).

These two examples seem to indicate that once the PNG students were asked to operate with numbers beyond mechanically manipulating whole numbers, they were in difficulty. Even with whole numbers there is some concern. An item on the test read 'Jill has to collect 200 coconuts. She has collected 155 coconuts and taken them away. How many more has she to collect?' A number of students read this item correctly and articulated an appropriate understanding of what the end point was. Some students then said one needed to subtract and proceeded to add, or said that adding was necessary and proceeded to, thus making a process skill or transformation error with an answer of 365. The author questioned a number of these students, asking them again what type of answer was required. Most correctly articulated the end point a second time and were still happy with an answer of 365. A few students saw there was a discrepancy but became bewildered and could not proceed. A similar result occurred for another item which asked how much five cups would hold if each contained 230 ml. A typical answer was 46 ml, with a number of students unaware of an inconsistency between the numbers. It seemed reasonably clear that an informal comparison of numbers was not carried out in any meaningful way by these students. They did not have that somewhat vague but important concept of 'number sense'.

There was some rise in the frequency of encoding errors. All the items which comprised this test were of an extended answer type. In finishing each one, the student was expected to take the mathematical result from a calculation and write it in a space provided for the answer. When doing this for a number of items, it was expected that a unit or an appropriate designating word such as 'coconuts' would be added. This encoding transformation is a language-based skill. A particularly frequent mistake was the inappropriate use of money units. The correct answer for one item was 29t: answers such as .29t, 29 or K.29t were common. Another common error was the omission of the decimal point. In the resultant probing by the interviewer, it became clear that the student did not regard this 'full stop' as an integral part of the number, confirming the observations made with respect to the comprehension errors.

In considering the language errors that were made, it was considered whether any of these errors occurred because of an interaction between the students' mother tongue and English, the language used for the test. An examination of the words that were not known or had been badly mispronounced, which had been noted as reading errors, was made. This was followed by listing particular words and phrases which had been frequently noted as specific comprehension errors. There did not seem to be any reason to believe that words from the mother tongue were causing these errors. Many of the words were specifically mathematical such as 'sum' and 'multiply', or more common words given a specific mathematical meaning in this context, such as 'difference' and 'shared'. However, it may be that this rather surface examination of the possible effect of the mother tongue has not detected an underlying influence. Such an influence may not be at all obvious for students who are at least 11 years old and have experienced 6 years of schooling.

Practical implications

The findings discussed above have direct implications for classroom practices, and one of the uses to which they have been put is in-service work with teachers. It is instructive to consider some of the outcomes of such exercises.

First, there was some surprise among teachers that the word problems are as difficult as the data suggest, and that some of that difficulty is language related. Good discussions on the need for a wide repertoire of mathematical words to fully understand and describe mathematical processes have resulted. It has been suggested that such vocabulary needs to be taught in the context of solving problems, and taught as an isolated topic as used to occur, and sometimes still does, in language classes. In such in-service sessions, it has been noted how the continued misuse of particular symbol or terms, like the decimal point, may indicate that a student has little knowledge of a basic mathematical concept, such as place value.

Secondly, some teachers attending in-service sessions came to see the hierarchy proposed by Newman as a useful approach in leading students through a problem, rather than just telling them what to do. This point seems to be rather more critical in PNG than in Western countries. Although various styles of teaching, other than a very teacher-centred one, have been espoused in PNG from time to time, they have had little or no impact throughout the system. Most teachers have had a minimal amount of pre-service training, and the in-service training available has been spasmodic. Hence, teachers have tended to teach how they were taught, and in a mode which is safe for them and ensures their authority is not challenged. Such an approach to teaching also seems to be reinforced by the traditional cultures of the land where knowledge is 'handed down', not created by each individual for him or herself. Hence, when a student cannot solve a problem, the 'natural' action of most PNG teachers is to 'do it' for the student. The solution is then copied out.

But for those teachers who wished to try another way of teaching and were in attendance at in-service sessions where the above studies were discussed, the Newman model offered an easily adapted and relatively simple model to start with. Instead of just showing what to do, teachers were encouraged to first ask the student to 'read the question out loud'. Any words not known or mispronounced could then be discussed. It was then suggested that teachers should then leave the student to make another attempt on the problem. If, on checking with the student later it was clear little progress had been made, it was suggested that teachers should ask the student 'what is the question asking you to do [or to find out]'. When the end point of the problem was articulated in discussion, any special symbols or terms were checked. Again, the teacher would leave the student so that another attempt could be made. This procedure could be worked through step by step following the hierarchy. No formal follow-up study on teachers' attempts at implementing this teaching strategy has been possible, but anecdotal evidence from a number of teachers gives encouragement. The simplicity and direct approach of the model seems to be of great benefit in PNG.

Conclusion

These studies have shown again that the solving of word problems is a multi-step process during which errors can occur at any step. It would appear that PNG students have particular difficulty with reading and comprehending word problems. They also have a limited mathematical vocabulary. The work with teachers indicates that systematic attempts to tackle these difficulties, based on increased awareness of the linguistic issues and exploitation of the Newman model, are promising.

One further point which can be made about the present studies is the role of non-cognitive variables. In the first study, it was noted that 'misplaced confidence' seemed to be one variable which correlated significantly with the frequency of reading errors. Whether non-cognitive variables play a role in explaining why PNG students seemed to make more language-based errors than native English speakers, or whether it is due more to the bilingual context they are in, or whether it is an outcome of the style of teaching that PNG students experience is an open question. However, it has been noted elsewhere that in some respect PNG students have a different profile on non-cognitive measures to Western students (see Clarkson, 1984b; Clarkson and Leder, 1984). It would not surprise me if they do indeed play a significant role. What it is, and how it interacts with teaching styles and the bilingual context will take some time to tease out.

Above all, it should be stressed that the problems of mathematical education in multilingual societies are not to be seen as insuperable. As has been seen, there are bases and strategies for effective intervention. And there are grounds for optimism in the achievements of individuals who have progressed despite these hurdles. Atawe, with whom this chapter began, is a case in point. Going to school, holding pencils, doing drawings, and so on was so fascinating for him that there were no complaints. From that type of start in the 1960s, Atawe took out a degree majoring in mathematics, and for some years was a staff colleague of the author while employed at the Papua New Guinea University of Technology. Atawe's start was not unusual for a boy who was fortunate enough to have the opportunity to go to school in PNG in the 1960s. His success in the education system was not a typical experience, but it is the kind of outcome that could be promoted with a more focused attention on the specific challenges of learning mathematics in a multilingual society.

Further reading

Clarkson, P.C. (1983). Types of errors made by Papua New Guinean students. *Educational Studies in Mathematics*, **14**, 355–67.

Clarkson, P.C. and Leder, G.C. (1984). Causal attributions for success and failure in mathematics: A cross-cultural perspective. *Educational Studies in Mathematics*, **15**, 413–22.

Newman, M.A. (1977). An analysis of sixth grade pupils' errors on written mathematical tasks. In M.A. Clements and J. Foyster (eds), *Research in mathematics education in Australia, 1977*, Vol. 1. Melbourne: Swinburne Press.

21 Number experience and performance in Australian Aboriginal and Western children

Judith Kearins

Is number learning the same for children of all cultures?

Health, education and police workers of ordinary Western groups use a number of questions on their first official contact with individuals, the answers to which they see as providing basic data necessary to the social definition of each person. Questions asked of young and old always include requests for name and age. Failure to provide an answer to one of these questions may be attributed to shyness or unwillingness to cooperate, but rarely to ignorance. Yet some people may be unable to say how old they are, and some may not understand the question, nor what form the answer should take. This has been the case for Australian Aboriginal people in the past, and may still be so for some older Aboriginal people (as it has been for traditional hunter-gatherer people elsewhere).

The idea of assigning changing numbers to each person over time, which is what accurate age knowledge entails, can be expected to be mysterious and irrelevant for people who have no tradition of measurement. A single number may be difficult to remember in isolation (as a bank or tax file number may be for Westerners), while its change over time would require full familiarity with the Western calendar and counting system. Since hunting and gathering people, among others, did not in the past and may not now have such familiarity, they may not be able to say how old they are.

For Westerners the story is rather different. The measurement of age, with physical change over time, is accepted without question as a personal datum of significance, important to self-definition and to the descriptions of others. [This may not have been so for very long. Aries (1962) considers that five centuries ago only a few exceptionally well-educated people would have known their ages exactly. Even so, the commonly used liturgical calendar would have made for fairly accurate estimations.]

In this chapter, I will consider the context and consequences of different cultures' practices with respect to number learning with particular reference to the contrast between Western and Aboriginal people. I begin with an outline of the evolution of different uses of number systems by Westerners and Aboriginals, showing the relationship between ecological adaptation and socialization practices. Then, I describe studies of my own comparing the early number uses of Aboriginal and Anglo-Australian children, and number knowledge, measurement and direction in older children. These studies show differences in orientation and performance with respect to basic mathematical skills, and I consider their implications for mathematical education.

The evolution of number systems in Western and Aboriginal societies

Present-day Western children accept as a matter of course the everyday use of number, so that the division of time, growth, area, distance and most of life's necessities into numerical units seems natural. This could not happen unless their culture, through their parents or guardians, encouraged considerable early learning of number. Since cultures provide 'a recipe for learning' (Irvine, 1988: 166), and each culture demands, encourages and rewards somewhat different collections of cognitive skills, it has to be assumed that number learning has come to be seen as important by Western parents, but not necessarily by parents of other cultural groups. Western parents apparently assume, for instance, that very young children should be introduced to counting, since they use deliberate teaching of number at least from a child's second year (Durkin *et al.*, 1986). Before this, their use of number in child rearing may reflect their own orientations rather than intention to teach, but it has the effect of introducing the concept and vocabulary of number to infants. Counters are strung across cribs as playthings, for instance, the abacus is given as an early toy, and we have many counting rhymes and games which we begin to play with children in infancy. These include baby tossing to the count of 'one, two, three . . .'; toe and finger play; games such as 'Round and round the garden'; rhymes such as 'One, two, buckle my shoe', 'One, two, three, four, five, once I caught a fish alive'; songs such as 'Green grow the rushes-O'; and attention to birthdays. As Sinclair points out (this volume), written numbers are pervasive in the Western child's everyday environment.

Partitioning by number in Western culture has a long developmental history (of many thousands of years: Eves, 1976), so that its importance to Western parents cannot be surprising. The earliest surviving record relating to number is of tallying (before the development of numerals) on an 8000-year-old bone found at Ishango in Zaire, which shows numbers preserved by notches cut in the bone shaft (Eves, 1976; Bynum *et al.*, 1981). Historians assume that most tally records would have been more ephemeral than notches in bone (e.g. piles of pebbles or sticks, or notches in wood), so that we have no idea how old the practice may have been. Later, word tallies may have developed, then symbols to stand for number words, while with more extensive counting would have come systematization into groups according to a scale or base (such as base 10). The numeral or written record goes back only as far as the first dynasty (3400 B.C.) in Egypt, although arithmetical knowledge and computation can be inferred for the Sumerians as far back as 5000 B.C. (Eves, 1976). Our own number system with base 10 is the Hindu-Arabic system, invented by the Hindus and brought to Europe by the Arabs, probably as 'traders and travellers of the Mediterranean coast' (Eves, 1976: 19).

Most historians (e.g. Bynum *et al.*, 1981; the *Encyclopaedia Britannica*) are in essential agreement with an agricultural or pastoral view of the origin of number systems. According to the *Encyclopaedia Britannica* (1963: 610–12), the pure decimal system may have arisen with nomadic herders, who needed high numbers for counting flocks of camels, sheep, cows and horses, and from them the practice spread elsewhere. Bynum *et al.* (1981: 249) say that from the early reckoning on fingers, pebbles and so on came 'a calendar in connection with agriculture'. Hogben (1960: 29) has argued that 'to regulate a seasonal economy a solar calendar is an imperative requirement', and that only in a settled lifestyle can seasonal variations be accurately reckoned. (He might have added that such reckoning is more likely to begin in latitudes far from the equator, such as northern Europe, where seasonal variations are more marked and growing seasons shorter than in tropical and subtropical regions like Australia.) Hogben goes on to point out that days occur, and must be counted, in sequence rather than all together, and that early numerical records must have involved the recording of time (through marks chipped on a tree trunk or pillar, perhaps) for the seasonal predictions essential to agriculturists.

Other measurements of importance to agricultural people may have gradually developed, either at the same time as the solar calendar or later. The approximation of distances, for instance, from one settlement to another, or from settlement to summer pasture in snow country, must have occurred, although this would have been initially in terms of sun time rather than scale measurement, and may have remained so until relatively recent times. The estimation of food supplies would have needed greater formality, especially in a cold country, since short growing seasons involved the storage over the rest of the year, and particularly the winter, of enough food for the people and livestock of each settlement. This, and the number reckoning necessary in animal husbandry, meant that settlement inhabitants would need to develop a variety of measurement skills. Since children must be shaped from earliest childhood towards the sort of adult behaviour each society requires, number would presumably be taught to the very young, or learned by children observing the activities of their elders and absorbing what was important to these adult role models. The current Western teaching of number in early childhood presumably stems from such early beginnings, and seems to have become such an accepted practice of child rearing that those practising it may consider that all others do so too.

In hunting and gathering groups, there would never have arisen a need for measurement such as that of the settled pastoral/agricultural world. In this lifestyle, animals are not kept for food, nor are crops grown. Since food grows naturally or is present in animal form, over a wide area and all seasons, little storage against arid seasons occurs. This would, in any case, have been neither likely nor necessary for any but arctic hunters, since most hunter-gatherer people lived in temperate or tropical zones. In Australia, for instance, most food kept in summer without refrigeration spoils because of the heat; and winter, rather than being a desolate season as in northern Europe, tends to be a time of relative plenty. In all but drought-affected times, the Australian Aboriginal diet was rich and varied, and food plentiful within a short working day (depending somewhat on locality and season; compare the 2.5-day working week of the Kalahari people: Lee, 1968). Further, it is unlikely that accurate time measures would have developed in tropical and temperate zones without an economic spur (present in Egypt and Mesopotamia, for example), since variation in daylength and sun position is far less obvious than in northern Europe. While names for different seasons occur in Aboriginal languages, little or no economic dependence on greater, or different, economic activity would occur with the onset of any season and, beyond day and night, small time units did not need to be recognized. Age, size, development and behaviour signalled the time for progress into adulthood (as they probably do for all people). This means that neither economic nor latitudinal constraints or requirements would have made likely the development of complex counting. This would not have been needed, since trade was minimal and barter was carried out through group representatives. In ritual, counting was not used, nor in warfare, which seems to occur in organized ways only among settled people.

Although a few number names exist in Aboriginal languages, though these vary with language group (Harris, 1987, reports words to 10, routinely, with precise number names for 20, 30, 40 in at least one case, while names for 100 and 1000 may have occurred in the Torres Straits Islands), it seems to have been the case that counting had little importance, as must also have been the case for other hunting and gathering people (see e.g. Blurton Jones and Konner, 1976). Under these circumstances, it cannot be expected that traditions for the teaching of counting to children would have developed in hunter-gatherer groups. It may also be the case that number learning needs a certain amount of deliberate and verbal teaching, and that it is therefore most likely in cultures whose members shape young children by largely verbal means. While this occurs for Westerners, it seems not to be the case for hunter-gatherers, where rearing for independence rather than compliance (Barry *et al.*, 1959) involves high child autonomy from infancy (e.g. Hamilton 1981) and an absence of the requirement that children must listen, attend or respond to their parents or other adults (e.g. Kearins, 1985).

Number knowledge in Australian Aboriginal and non-Aboriginal children

Given the large differences in number teaching and the emphasis likely between Westerners and hunter-gatherers, then, it should be expected that Anglo-

Australian and Australian Aboriginal children will differ in number knowledge, especially if the Aboriginal children have been reared in traditional ways or by practices stemming from these ways.

In an investigation of preschool children in the Perth metropolitan area of Western Australia, Judith Butters and I (Kearins and Butters, 1986) compared early number use in Aboriginal and Anglo-Australian children aged between 4 years and 4 years 6 months. The Aboriginal children ($N=35$) attended one of two special 'pre-preschools' designed to introduce them to Western school ways, while the Anglo-Australian children attended either the kindergarten at the University of Western Australia ($N=19$), where they can be expected to have come from fairly advantaged backgrounds, or a day-care centre in a less affluent suburb, where their backgrounds were varied, but where they were more likely to have come from the lower end of the socio-economic range. A small group of remote area Aboriginal children ($N=14$) was also asked counting and age questions by Judith Butters.

We found marked differences in age knowledge (i.e. knowing the correct number for their age in years) between Aboriginal and Anglo-Australian children. All 19 of the day-care children (100%) and all but one of the university children (95%) gave their correct ages, but only 16 (about 46%) of the urban Aboriginal children, and 4 (about 28%) of the remote area Aboriginal children did so.

When we asked children to count for us, the university children counted, on average, to about 16 (note that this is consistent with the observation made by Fuson, this volume, for middle-class Western children of this age). The day-care children counted only about half as far (mean 8.65, median 7), while urban Aboriginal children were not far below this (mean 6.95, median 6). The remote area Aboriginal children fell a little below this again (mean 6.2, median 4.5).

From these enquiries, then, it looks as though age knowledge may be acquired very early by young Anglo-Australian children, but not necessarily by Aboriginal children of the same age even if they live in a metropolitan region. Counting knowledge in this study, on the other hand, was relatively higher in Anglo-Australian children of higher socio-economic status (as would be expected from class-difference reports in the literature), but not very different for the less advantaged Anglo-Australian children and the Aboriginal children from both the urban and remote areas.

In a later study, I looked at number knowledge in older urban Aboriginal and non-Aboriginal children, and also in children of the same ages at a remote north-western school (Kearins, 1989). In this study, children in grade 2 (aged 7–8 years), grade 5 (aged 10–11 years) and junior high school (14–17 years, median approximately 14.5 years) were asked their ages, and distances to either Perth city (metropolitan children) or the nearest large town (north-west children). All of the urban children, both non-Aboriginal ($N=57$) and Aboriginal ($N=46$), gave their correct ages, as did a small ($N=15$) mixed-age group of north-west non-Aboriginal children (grades 2–6). For Aboriginal children of the north-west, however, age knowledge was not so certain; 46.5% of grade 2 ($N=17$), 68.2% of grade 5 ($N=22$) and 57.2% of high school ($N=18$) children gave their correct ages. When the scores were examined for accuracy to within 1 year (e.g. a child might give 8, or 6, as her age when it was actually 7), the figures improved markedly to 78.6, 100 and 93% for grades 2, 5 and high school students, respectively. This must have occurred because correct-to-the-day age was not seen as important by Aboriginally oriented children even if their parents, and, in many cases grandparents, had some Western schooling, and although almost all lived in a modern town. These children tended to say they were 9, for instance, from the beginning of a year in which they turned 9, or to continue to give their age as 13, for instance, although they had turned 14. Younger children had considerably more difficulty with personal number data such as age and grade level, since some gave their age as '2' when this was their grade level rather than their age. These instances suggest that day-by-day number had not been used with familiarity by these children.

Distance estimation showed, again, that measurement seemed to be less important to Aboriginal than to non-Aboriginal children, for whom age improvement was also marked. The scoring criterion in this case was a fairly liberal one, children unable to estimate distance being asked for a travel time estimate instead. Urban non-Aboriginal accuracy ranged from 0% at grade 2, 31% at grade 5 to 100% at high school level. Urban Aboriginal high school children made only 44% accurate estimations, however, and 11% and 0% at grades 5 and 2, respectively. North-west Aboriginal children of high school age were approximately as accurate as the urban Aboriginal group (43%), although the two younger age groups were somewhat more accurate than urban children (35% at grade 5, a little ahead even of the urban non-Aboriginal children; and 14% at grade 2, better than both urban Aboriginal and non-Aboriginal grade 2 children). Distance knowledge is important to car drivers in remote areas since fuel consumption is always tied to distance. Almost all families of the children interviewed drove cars, but most of these children said that they ran out of fuel 'lots of times', and no child was without such experience. Spare fuel cans were carried by all to help avoid this situation, although not always with success.

As part of both the above studies children were asked, in addition to measurement questions, to make direction judgements. Here, the story was quite different, Aboriginal children of both urban and remote areas showing high skill relative to that of the non-Aboriginal children. In the preschool study, for instance, no Anglo-Australian of the university kindergarten, and only one day-care child, could indicate their home direction accurately (i.e. to within 45 degrees), whereas 58% of the Aborigines did so, although most lived farther from the preschool centres. Among the older children of the second study, grade 2 Aboriginal children showed greater accuracy than non-Aboriginal high school students – who were, as would be expected, the best performers of the non-Aboriginal group. The distance over which direction indications were made seemed to be unimportant. Remote area children indicated the direction of a town 359 km away, while urban children indicated the direction of Perth city, 12–15 km distant. Despite this difference, the mean error shown by both groups of Aboriginal children shows a very similar profile across age levels. For grades 2, 5 and high school urban groups, the mean direction error was approximately 25, 20 and 9 degrees, while for remote area groups it was approximately 18, 14 and 12 degrees. This compares with the mean errors for urban non-Aboriginal groups of 63, 52 and 27 degrees. That rural living does not necessarily affect direction skill was indicated by the large error (57 degrees) made by the mixed-age non-Aboriginal group of the north-west school. Children brought up on pastoral stations or large farms in remote areas may make smaller errors, since their parents would presumably impress upon them the dangers of not knowing the way; such a group was not available in this case, however, although a few of the remote area non-Aboriginal group were station or non-town-living children.

These studies might promote an appreciation of the effect on child learning of different cultural backgrounds. In the present case, number use could be expected to differ in importance for the forebears of the two child groups assessed, such use having developed from an agricultural, but not a hunter-gatherer, background. Children of Western and Australian Aboriginal backgrounds conformed to these expectations in their number use and apparent familiarity with number. Even urban Aboriginal children showed approximately the remote area patterns (although age knowledge characterized all urban but not remote area children of grade 2 and above). The urban/remote area similarities, which extend to the estimation of direction – clearly a skill of great survival value to nomadic people – suggest that the type of learning acquired in early childhood does not necessarily change through adult contact with a different culture. This may be so even when the new culture is dominant, and contact is high, as might be expected in urban living.

Practical implications

The formal education of Australian Aboriginal children is likely to suffer as a consequence of these differences, even though they are predictable from the widely divergent historical backgrounds of the two groups in question. Aboriginal children who begin school not knowing much about number, and apparently not considering it important, may be seen by teachers as less intelligent, or less capable of learning, than the more familiar Western children for whom number is already important.

Aboriginal children may indeed have great difficulty in number learning, since early acquisition of number concepts may facilitate later number learning. Such facilitation should occur for Western children, whose early everyday number use within their families will have taught crucial number relationships and rules such as sequential invariance, as well as common number and measurement usage. Unfortunately, this cultural learning factor can be expected to work against Aboriginal children, since teachers are unlikely to see basic number as needing to be taught if most children beginning school are already equipped with this knowledge. Yet for Aboriginal children, lacking the background of gradual but quite intensive number learning which seems to be common to most Westerners, basic number teaching is clearly necessary. Without it, children are likely to find number work confusing and difficult and, realistically, to anticipate failure in mathematics. Furthermore, while success is positively reinforcing and is likely to lead to interest in further learning, failure is not, which means that Aboriginal children are doubly disadvantaged.

Aboriginal child rearing, as earlier noted, differs markedly from the Western pattern, and this difference must lead to some variation in learning style. The relative autonomy enjoyed from infancy by young Aboriginal children gives them freedom to learn by visual means and to become skilled observers, whereas Western children, seen by their parents as necessarily dependent, do not have this freedom. As a result, the latter children are more likely to develop verbal skills (of both listening and speaking) and to develop relatively verbal habits of learning. While it should be the case that the cognitive strengths of children of all cultures are able to be used in school, educational instruction within any system favours the children of that system. Major Western and Eastern educational systems expect the learning, and teach the material, seen as important within their own cultures, and cannot easily accommodate the learning or interests of children from other cultures. In Australia, this means that Aboriginal children are mostly unable to use their outside learning in school. With informed and sensitive teachers, however, and especially in school where Aboriginal children constitute majority populations, they should be able to learn by visual as well as verbal means, instead of by the predominantly verbal means mostly used by Westerners.

The known cognitive strengths of Aboriginal children (in, for example, visual spatial memory, physical skill, direction finding skill, wildlife knowledge: Kearins, 1983; 1986; 1989) are not often able to be used directly in the Western school setting. Apart from practical problems, many teachers may have difficulty in accepting the notion that cultural differences are likely in the patterns of ability which children bring to school, even though, as with Australian Aboriginal children, nurture rather than nature must be seen to be responsible. Strong belief that all children are alike and therefore must be taught in the same way, even in early school grades, can only disadvantage children whose early learning does not match the Western teacher's expectations. If differences in patterns of knowledge were known to, or accepted by, teachers of early grades, however, the bicultural learning which Aboriginal children need overtly to develop might become more possible to them through schooling. If this happened, special school provision for mathematics might become unnecessary by the later grades, since children would have had special teaching in the earlier grades. Advantage might then more readily be taken of Aboriginal knowledge (in environmental and wildlife areas for example) in an education system which became more relaxed about the basic school learning success of Aboriginal children.

Further reading

Kearins, J. (1981). Visual spatial memory in Australian Aboriginal children of desert regions. *Cognitive Psychology*, **13**, 434–60.

Kearins J. (1985). Cross-cultural misunderstandings in education. In J. Pride (ed.) *Cross-cultural encounters: Communication and miscommunication*. Melbourne: River Seine.

Bibliography

Adda, J. (1982). Difficulties with mathematical symbolism: Synonymy and homonymy. *Visible Language*, **16**, 205–14.

Agnoli, F. and Zhu, J. (1989). One, due, san: Learning to count in English, Italian, and Chinese. *Proceedings of the Biennial Meeting of the Society for Research in Child Development*, **6**, 75 (abstract).

Aiken, L.R. (1972). Language factors in learning mathematics. *Review of Educational Research*, **42**, 359–85.

Allardice, B.S. (1977). The development of written representations for some mathematical concepts. *Journal of Children's Mathematical Behavior*, **1**, 135–48.

Anghileri, J. (1988). *An investigation of young children's understanding of multiplication.* Unpublished Ph.D. dissertation. King's College, University of London.

Aries, P. (1962). *Centuries of childhood.* London: Jonathan Cape.

Atkinson, P. (1981). Inspecting classroom talk. In C. Adelman (ed.), *Uttering, muttering.* London: Grant McIntyre.

Austin, G.F. (1975). Knowledge of special concepts obtained by an adolescent deaf population. *American Annals of the Deaf*, **120**(3), 360–70.

Austin, J.L. and Howson, A.G. (1979). Language and mathematical education. *Educational Studies in Mathematics*, **10**, 161–97.

Avesar, C. and Dickerson, D.J. (1987). Children's judgements of relative number by one-to-one correspondence: A planning perspective. *Journal of Experimental Child Psychology*, **44**, 236–54.

Balacheff, N. (1988). Aspects of proof in pupils' practice of school mathematics. In D. Pimm (ed.), *Mathematics, teachers and children.* Sevenoaks: Hodder and Stoughton.

Barham, J.C. (1987). Giving deaf children the language of maths. *Special Children*, **14**, 10–12.

Barham, J.C. (1988). *Teaching mathematics to deaf children.* Unpublished Ph.D. dissertation, University of Cambridge.

Barham, J.C. and Hodge, M. (1988). Animated mathematics – a suite of programs for the Spectrum computer. *Microscope*, **23**, 28–31.

Barnes, D. (1976). *From communication to curriculum.* Harmondsworth: Penguin.

Baroody, A.J. (1987). The development of counting strategies for single-digit addition. *Journal for Research in Mathematics Education*, **18**, 141–57.

Baroody, A.J. and Ginsburg, H.P. (1983). The effects of instruction on children's understanding of the 'equals' sign. *Elementary School Journal*, **84**, 199–212.

Baroody, A.J. and Ginsburg, H.P. (1986). The relationship between initial meaningful and mechanical knowledge of arithmetic. In J. Heibert (ed.), *Conceptual and procedural knowledge: The case of mathematics*. Hillsdale, N.J.: Lawrence Erlbaum Associates Inc.

Barry, H., Child, M. and Bacon, I. (1959). Relation of child training to subsistence economy. *American Anthropologist*, **61**, 51–63

Bednarz, N. and Janvier, B. (1982). The understanding of numeration in primary school. *Educational Studies in Mathematics*, **13**, 33–57.

Beeney, R., Jarvis, M., Tahta, D., Warwick, J. and White, D. (1982). *Geometric images*. Derby: ATM.

Beilin, H. (1975). *Studies in the cognitive basis of language development*. London: Academic Press.

Bell, A. (1984). Short and long term learning – experiments in diagnostic reaching design. *Proceedings of the Eighth International Conference for the Psychology of Mathematics Education*. Sydney: International Group for the Psychology of Mathematics Education.

Bell, A., Swan, M. and Taylor, G. (1981). Choice of operations in verbal problems with decimal numbers. *Educational Studies in Mathematics*, **12**, 399–420.

Bennett, D.C. (1975). *Spatial and temporal uses of English prepositions: An essay in stratificational semantics*. London: Longman.

Bilsky, L.H., Blachman, S., Chi, C., Mui, A.T. and Winter, P. (1986). Comprehension strategies in math problem and story contexts. *Cognition and Instruction*, **3**, 9–126.

Bishop, A.J. (1980). Spatial abilities and mathematics education – a review. *Educational Studies in Mathematics*. **11**(3), 257–69.

Bishop, A.J. (1983). Space geometry. In R. Lesh and M. Landau (eds), *Acquisition of mathematics concepts and processes*. London: Academic Press.

Bishop, A.J. (1985). The social construction of meaning – a significant development for mathematics education? *For the Learning of Mathematics*, **5**, 24–8.

Bishop, D.V.M. (1987). The concept of comprehension in language disorder. *Proceedings of First International Symposium of AFASIC*. London: AFASIC.

Black, M. (1979). More about metaphor. In A. Ortony (ed.), *Metaphor and thought*, Cambridge: Cambridge University Press.

Blundell, D. (1988). *The language of problem solving and Logo*. Unpublished M. Phil. thesis, University of Cambridge.

Blurton Jones, N.G. and Konner, M. (1978). Kung knowledge of animal behaviour. In R.B. Lee and I. de Vore (eds), *Kalahari hunter-gatherers*. Cambridge, Mass.: Harvard University Press.

Boole, M.E. (1931). *Lectures in mathematics education*. London: University of London Press.

Brainerd, C.J. (1973a). Mathematical and behavioral foundations of number. *Journal of General Psychology*, **88**, 221–81.

Brainerd, C.J. (1973b). The origins of number concepts. *Scientific American*, **228**, 101–109.

Brainerd, C.J. (1979). *The origins of the number concept*. New York: Praeger.

Briars, D.J. and Larkin, J.H. (1984). An integrated model of skills in solving elementary word problems. *Cognition and Instruction*, **1**, 245–96.

Briars, D. and Siegler, R.S. (1984). A featural analysis of preschoolers' counting knowledge. *Developmental Psychology*, **20**, 607–618.

Brissenden, T.H.F. (1972). Deciding the objectives of primary mathematics. In M Chazan (ed.), *Aspects of primary education.* Cardiff: University of Wales Press.
Brissenden, T. (1985). Rewriting the lesson script. *Mathematics in School*, **14**, 2–6.
Broadbent, F.W. and Daniele, V.A. (1982). A review of research on mathematics and deafness. *Directions* (*Gallaudet College*), **3**(1), 27–36.
Brown, G. (1982). The spoken language. In R. Carter (ed.), *Linguistics and the teacher* London: Routledge and Kegan Paul.
Brown, M. (1981). Number operations. In K. Hart (ed.), *Children's understanding of mathematics: 11–16.* London: John Murray.
Bruner, J.S. (1983). *Child's talk: Learning to use language.* Oxford: Oxford University Press.
Bryant, P.E. (1974). *Perception and understanding in young children: An experimental approach.* London: Metheun.
Bryant, P.E. (1985). The distinction between knowing when to do a sum and knowing how to do it. *Educational Psychology.* **5**(3, 4), 207–216.
Bynum, W.F., Browne, E.J. and Porter, R. (eds), (1981). *Dictionary of the history of science.* London: Macmillan.
Carpenter, T.P. (1975). Measurement concepts of first- and second-grade students *Journal of Research in Mathematics Education*, **6**, 3–14.
Carpenter, T.P. (1985). Learning to add and subtract: An exercise in problem solving. In E. Silver (ed.), *Problem solving: Multiple research perspectives.* Philadelphia: Franklin Institute Press.
Carpenter, T.P. and Moser, J.M. (1983). The acquisition of addition and subtraction concepts. In R. Lesh and M. Landau (eds), *Acquisition of mathematics: Concepts and processes.* London: Academic Press.
Carpenter, T.P. and Moser, J.M. (1984). The acquisition of addition and subtraction concepts in grades one through three. *Journal for Research in Mathematics Education*, **15**, 179–202.
Carpenter, T.P., Hiebert, J. and Moser, M. (1981). Problem structure and first grade children's initial solution processes for simple addition and subtraction problems. *Journal for Research in Mathematics Education*, **12**, 27–39.
Casey, D.P. (1979). *An analysis of errors made by junior secondary pupils on written mathematical tasks.* Unpublished M. Ed. thesis, Monash University.
Cauley, K.M. (1988). Construction of logical knowledge: Study of borrowing in subtraction. *Journal of Educational Psychology*, **80**(2), 202–205.
Clarkson, P.C. (1980). The Newman error analysis – some extensions. In B.A. Foster (ed.), *Research in mathematics education in Australia, 1980*, Vol. 1. Hobart: Mathematics Education Research Group of Australia.
Clarkson, P.C. (1983). Types of errors made by Papua New Guinean students. *Educational Studies in Mathematics*, **14**, 355–67.
Clarkson, P.C (1984a). Word problems at year 5 level. In P.C. Clarkson (ed.), *Research in mathematics education in Papua New Guinea, 1984.* Lae: Mathematics Education Centre, Papua New Guinea University of Technology (ERIC, ED 251 324).
Clarkson, P.C. (1984b). Papua New Guinean students' perceptions of mathematics lecturers. *Journal of Educational Psychology*, **76**(60), 1386–95.
Clarkson, P.C. (1986). On consistency of classifying errors. *Research in Mathematics Education in Australia*, December, 22–4.
Clarkson, P.C. and Leder, G.C. (1984). Causal attributions for success and failure in mathematics: A cross cultural perspective. *Educational Studies in Mathematics*, **15**, 413–22.

Clements, M.A. (1980). Analysing children's errors on written mathematical tasks. *Educational Studies in Mathematics*, **11**, 1–12.

Clements, M.A. (1982). Careless errors made by sixth-grade children on written mathematical tasks. *Journal for Research in Mathematics Education*, **13**, 136–44.

Clements, M.A. and Lean, G.A. (1981). Influences on mathematics learning in community schools in Papua New Guinea: Some cross-cultural perspectives. *Mathematics Education Centre Report No. 13*, Papua New Guinea University of Technology.

Cobb, P. (1987). An investigation of young children's academic arithmetic contexts. *Educational Studies in Mathematics*, **18**, 109–124.

Cockcroft, W. (1982). *Mathematics counts*, London: HMSO.

Corran, G. and Walkerdine, V. (1981). *The practice of reason*. Project Report – Cognitive Development and Educational Practice: Pupil Progress in Primary School Mathematics, University of London.

Cowan, R. (1984). Children's relative number judgements: One-to-one correspondence, recognition of noncorrespondence, and the influence of cue conflict. *Journal of Experimental Child Psychology*, **38**, 515–32.

Cowan, R. (1987a). When do children trust counting as a basis for relative number judgements? *Journal of Experimental Child Psychology*, **43**, 328–45.

Cowan, R. (1987b). Assessing children's understanding of one-to-one correspondence. *British Journal of Developmental Psychology*, **5**, 149–53.

Cowan, R. and Biddle, S. (1989). Children's understanding of one-to-one correspondence in the context of sharing. *Educational Psychology*, **9**, 133–40.

Cowan, R. and Daniels, H. (1989). Children's use of counting and guidelines in judging relative number. *British Journal of Educational Psychology*, **59**, 200–211.

Cross, G. and Morris, J. (1980). Linguistic feedback and maternal speech: Comparisons of mothers addressing infants, one-year-olds and two-year-olds. *First Language*, **1**, 98–121.

Dale, P.S. (1976). *Language development: Structure and function*, 2nd edn. New York: Holt, Rinehart and Winston.

Davis, A.M. and Lo, Y.Y. (1986). The effect of context on children's arithmetic. Unpublished manuscript, Department of Child Development, Institute of Education, London.

Davis, A.M., Bridges, A. and Brosgall, A. (1985). Showing how many: Young children's written representation of number. *Educational Psychology*, **5**(3, 4), 303–310.

Davis, R.B. (1984). *Learning mathematics: The cognitive science approach to mathematics education*. Norwood, N.J. Ablex.

Davydov, V.V. and Andronov, V.P. (1981). *Psychological conditions of the origination of ideal actions* (English translation), Project Paper 81–2. Madison, Wis.: Wisconsin Research and Development Center for Individualized Schooling.

Deane, P.D. (1988). Polysemy and cognition. *Lingua*, **75**, 325–61.

De Corte, E. and Verschaffel, L. (1985a). Beginning first graders' initial representation of arithmetic word problems. *Journal of Mathematical Behavior*, **4**, 3–21.

De Corte, E. and Verschaffel, L. (1985b). Working with simple word problems in early mathematics instruction. In L. Streefland (ed.), *Proceedings of the Ninth International Conference for the Psychology of Mathematics Education, Vol. 1: Individual contributions*. Utrecht: Research Group on Mathematics Education and Educational Computer Center, Subfaculty of Mathematics, University of Utrecht.

De Corte, E. and Verschaffel, L. (1987). The effect of semantic structure on first graders' solution strategies of elementary addition and subtraction word problems. *Journal for Research in Mathematics Education*, **18**, 363–81.

De Corte, E., Verschaffel, L. and De Win, L. (1985a). The influence of rewording verba problems on children's problem representations and solutions. *Journal of Educationa Psychology*, **77**, 460–70.

De Corte, E., Verschaffel, L., Janssens, V. and Joillet, L. (1985b). Teaching wor problems in the first grade: A confrontation of educational practice with results o recent research. In T.A. Romberg (ed.), *Using research in the professional life of mathematics teachers*. Madison: Center for Education Research, University of Wisconsin

Desforges, A. and Desforges, C. (1980). Number-based strategies of sharing in young children. *Educational Studies*, **6**, 97–109.

Dodwell, C.P. (1960). Children's understanding of number and related concepts *Canadian Journal of Psychology*, **14**, 191–205.

Doise, W.A. and Mugny, G. (1984). *The social development of the intellect*. Oxford: Pergamon Press.

Donaldson, M. (1978). *Children's minds*. London: Fontana.

Donaldson, M. (1982). Conservation: What is the question? *British Journal of Psychology*, **73**, 199–207.

Donaldson, M. and Wales, R.J. (1970). On the acquisition of some relational terms. In J.R. Hayes (ed.), *Cognition and the development of language*. New York: John Wiley.

Donaldson, M.L. (1986). *Children's explanations: a psycholinguistic study*. Cambridge: Cambridge University Press.

Donlan, C. and Hutt, E. (1989). *M.A.P: Mathematics Assessment Procedure for young language-impaired children*. London: I CAN.

Dore, J. (1979). Conversation and preschool language development. In P. Fletcher and M. Garman (eds), *Language acquisition*. Cambridge: Cambridge University Press.

Duncker, K. (1945). On problem solving. *Psychological Monographs*, **58**(5).

Durkin, K. (1981). Aspects of late language acquisition: School children's use and comprehension of prepositions. *First Language*, **2**, 47–59.

Durkin, K. (1986). Language and social cognition during the school years. In K. Durkin (ed.), *Language development in the school years*. London: Croom Helm.

Durkin, K. (1987). Minds and language: Social cognition, social interaction and the acquisition of language. *Mind and Language*, **2**, 105–140.

Durkin, K. (1988). The social nature of social development. In M. Hewstone, W. Stroebe, J.P. Codol and G.M. Stephenson (eds), *Introduction to social psychology: A European perspective*. Oxford: Blackwell.

Durkin, K. and Shire, B. (in press). Primary school children's interpretations of lexical ambiguity in mathematical descriptions. *Journal of Research in Reading*.

Durkin, K. Crowther, R.D. and Shire, B. (1986a). Children's processing of polysemous vocabulary in school. In K. Durkin (ed.), *Language development in the school years*. London: Croom Helm.

Durkin, K., Shire, B., Riem, R., Crowther, R. and Rutter, D.R. (1986b). The social and linguistic context of early number word use. *British Journal of Developmental Psychology*, **4**, 269–88.

Durkin, K., Manning J. and Romick, J. (submitted). Children's understanding of the word *big* in numerical contexts.

Earp, N.W. and Tanner, F.W. (1980). Mathematics and language. *The Arithmetic Teacher*, **28**(4), 32–4.

Edwards, A. and Bajpai, A. (1979). Preliminary report on the basic arithmetic test carried out in post-secondary institutions in Papua New Guinea in 1979. *Mathematics Education Centre Report No. 6.*, Papua New Guinea University of Technology.

Edwards, D. and Mercer, N. (1986). Context and continuity: Classroom discourse and the development of shared knowledge. In K. Durkin (ed.), *Language development in the school years*. London: Croom Helm.

Edwards, D. and Mercer, N. (1987). *Common knowledge: The development of understanding in the classroom*. London: Methuen.

Ellis, R. and Wells, G. (1980). Enabling factors in adult–child discourse. *First Language*, **1**, 46–62.

Eraut, M. and Hoyles, C. (1988). Groupwork with computers. *Journal of Computer Assisted Learning*, **5**,(1), 12–24.

Ericsson, K.A. (1975a). Instruction to verbalize as a means to study problem solving process with The 8-Puzzle: A preliminary study. Report of the Department of Psychology, University of Stockholm.

Ericsson, K.A. (1975b). Problem solving behaviour with The 8-Puzzle IV: Process in terms of sequences of moves. Report of the Department of Psychology, University of Stockholm.

Ericsson, K.A. and Simon, H.A. (1980). Verbal reports as data. *Psychological Review*, **87**(3), 215–51.

Eves, H. (1976). *An introduction to the history of mathematics*. New York: Holt, Rinehart and Winston.

Fauvel, J. (1987). Unit 6. IN MA290 *Topics in the history of mathematics*. Milton Keynes: Open University Press.

Fischbein, E., Deri, M., Sainati, Nello, M., and Sciolis, M. (1985). The role of implicit models in solving verbal problems in multiplication and division. *Journal for Research in Mathematics Education*, **16**,(1), 3–17.

Fisher, G. (1860). *The instructor – or young man's best company*. London: University of London.

Forman, E. (1988). The role of peer interaction in the social construction of mathematical knowledge. *International Journal of Educational Research*, **13**, 55–70.

Frydman, O. and Bryant, P.E. (1988). Sharing and the understanding of number equivalence by young children. *Cognitive Development*, **3**, 323–39.

Frye, D., Braisby, N., Lowe, J., Maroudas, C. and Nicholls, J. (1989). Young children's understanding of counting and cardinality. *Child Development*, **6**, 1158–71.

Furrow, D., Murray, P. and Furrow, M. (1985–6). Spatial term use and its relation to language function at two developmental stages. *First Language*, **6**, 40–51.

Furth, H.G. (1966). *Thinking without language*. New York: Free Press.

Fuson, K.C. (1986). Roles of representation and verbalization in the teaching of multi-digit addition and subtraction. *European Journal of Psychology of Education*, **1**, 35–56.

Fuson, K.C. (1988). *Children's counting and concepts of number*. New York: Springer-Verlag.

Fuson, K.C. (in press). Research on learning and teaching addition and subtraction of whole numbers. In G. Ledinhardt and R. Putnam (eds), *Cognitive research: Mathematics learning and instruction*.

Fuson, K.C. (in press). Conceptual structures for multiunit numbers: Implications for learning and teaching multidigit addition, subtraction, and place value. *Cognition and Instruction*.

Fuson, K.C. and Briars, D.J. (1990). Using a base-ten blocks learning/teaching approach for first and second-grade place-value and multidigit addition and subtraction. *Journal for Research in Mathematics Education*, **21**, 180–206.

Fuson, K.C. and Hall, J.W. (1983). The acquisition of early number word meanings. In H. Ginsburg (ed.), *The development of children's mathematical thinking*. London: Academic Press.

Fuson, K.C. and Kwon, Y. (in press). *Systems of number words and other cultural tools: Effects on children's early computations*. In J. Bideaval and C. Meljac (eds), Les chemins du nombre (Pathways to number). Villenevre d'Aseq, France: Presses Universitaires de Lille and Hillsdale, N.J.: Erlbaum.

Fuson, K.C. and Kwon, Y. (under review). *Korean children's understanding of multidigit addition and subtraction*. Manuscript under review.

Fuson, K.C., Richards, J. and Briars, D.J. (1982). The acquisition and elaboration of the number word sequence. In C. Brainerd (ed.), *Progress in cognitive development research, Vol 1: Children's logical and mathematical cognition*. New York: Springer-Verlag.

Fuson, K.C., Secada, W.G. and Hall, J.W. (1983). Matching, counting and conservation of number equivalence. *Child Development*, **54**, 91–7.

Fuson, K.C., Stigler, J.W. and Bartsch, K. (1988). Grade placement of addition and subtraction topic in China, Japan, the Soviet Union, Taiwan, and the United States. *Journal for Research in Mathematics Education*, **19**, 449–58.

Gardner, H., Winner, E., Bechofer, R. and Wolf, D. (1978). The development of figurative language. In K.E. Nelson (ed.), *Children's language*, Vol. 1. New York: Gardner Press.

Garofalo, J. (1987). Metacognition and school mathematics. *Arithmetic Teacher*, **35**, 22–3.

Gathercole, V.C. (1985). More and more and more about more. *Journal of Experimental Child Psychology*, **40**, 73–104.

Gelman, R. (1972). The nature and development of early number concepts. In H.W. Reese (ed.), *Advances in child development and behavior*, Vol. 7. London: Academic Press.

Gelman, R. (1982). Accessing one-to-one correspondence: Still another paper about conservation. *British Journal of Psychology*, **73**, 209–220.

Gelman, R. and Gallistel, C.R. (1978). *The child's understanding of number*. Cambridge, Mass.: Harvard University Press.

Gelman, R. and Meck, E. (1983). Preschoolers' counting: Principles before skill. *Cognition*, **13**, 343–59.

Gelman, R. and Meck, E. (1986). The notion of principle: The case of counting. In J. Hiebert (ed.), *Conceptual and procedural knowledge: The case of mathematics*. Hillsdale, N.J.: Lawrence Erlbaum Associates Inc.

Ginsburg, H. (1977). *Children's arithmetic: How they learn it and how you teach it*. Austin, Tex.: Pro-Ed.

Ginsburg, H.P. (1980). *Children's arithmetic: The learning process*, New York: Van Nostrand.

Ginsburg, H.P. and Russell, R.L. (1981). Social class and racial influences on early mathematical thinking. *Monographs of the Society for Research in Child Development*, **46**(6).

Glennon, V.J. (1981). *The mathematical education of exceptional children and youth*. Reston, Va.: National Council of Teachers of Maths.

Glucksberg, S., Hay, A. and Danks, J.H. (1976). Words in utterance contexts: Young children do not confuse the meanings of same and different. *Child Development*, **47**, 737–41.

Grauberg, E. (1985). Some problems in the early stages of teaching numbers to language-handicapped children. *Child Language Teaching and Therapy*, **1**(1), 17–29.

Gréco, P. (1962). Quantite et quotite. In P. Greco and A. Morf, *Structures numeriques elementaires*. Paris: Presses Universitaires de France.

Greig, J. (1858). *The young ladies guide to arithmetic*. London: University of London.
Griffiths, A.L. (1969). *Basic mathematics*. London: Oliver and Boyd.
Grober, E. (1976). Polysemy: Its implications for a psychological model of meaning. Unpublished Ph.D. dissertation, Johns Hopkins University.
Guitel, G. (1975). *Histoire comparee des numerations ecrites*. Paris: Flammarion.
HM Inspectors of Schools HMSO (1978). *Primary education in England*. London: HMSO.
Halliday, M.A.K. (1975). Some aspects of sociolinguistics. In *Interactions between language and mathematical education*. Copenhagen: UNESCO.
Halliday, M.A.K. (1978). *Language as social semiotic*. London: Edward Arnold.
Hamilton, A. (1981). *Nature and nurture*. Canberra: Australian Institute of Aboriginal Studies.
Hanley, A. (1978). Verbal mathematics. *Mathematics in School*, **7**(4), 27–30.
Harris, J. (1987). Australian Aboriginal and Islander mathematics. *Australian Aboriginal Studies*, **2**, 29–37.
Harris, M., Jones, D. and Grant, J. (1983). The nonverbal context of mothers' speech to infants. *First Language*, **4**, 21–30.
Hart, K. (ed.), (1981). *Children's understanding of mathematics: 11–16*. London: John Murray.
Harvey, R. (1983). I can keep going up if I want to – one way of looking at learning mathematics. In R. Harvey, D. Kerslake, H. Shuard and M. Torbe. *Mathematics, Language, Teaching and Learning* No. 6. London: Ward Lock.
Hatano, G. (1982). Learning to add and subtract: A Japanese perspective. In T.P. Carpenter, J.M. Moser and T.A. Romberg (eds), *Addition and subtraction: A cognitive perspective*, pp. 211–312. Hillsdale, N.J.: Lawrence Erlbaum Associates Inc.
Heinrichs, J., Yurko, D.S. and Hu, J.M. (1981). Two-digit number comparison: Use of place information. *Journal of Experimental Psychology: Human Perception and Performance*, **7**, 890–901.
Hiebert, J. (1982). The position of the unknown and children's solutions of verbal arithmetic problems. *Journal for Research in Mathematics Education*, **13**, 341–9.
Higgins, P. (1980). *Outsiders in a hearing world*. Beverley Hills, Calif.: Sage.
Hine, W.D. (1970). The attainments of children with partial hearing. *Journal of British Association of Teachers of the Deaf*, **68**(400), 129–35.
Hobart, D. (1980). Mind your language. *Child Education*, **51**, 7.
Hogben, L. (1960). *Mathematics in the making*. London: MacDonald.
Howard, C.F. (1887). *Art of reckoning*. London University of London Press.
Howell, A., Walker, R. and Fletcher, H. (1980). *Mathematics for schools, Fletcher Maths level 1*. London: Addison-Wesley.
Hoyles, C. (1985). What is the point of group discussion in mathematics? *Educational Studies in Mathematics*, **16**, 205–214.
Hoyles, C. (ed.) (1988). *Girls and computers*. Bedford Way Papers No. 34. London: Institute of Education.
Hoyles, C. and Noss, R. (1987). Children working in a structured Logo environment: From doing to understanding. *Recherches en Didactique de Mathematique*, **8**(12), 131–74.
Hoyles, C. and Sutherland, R. (1989). *Logo mathematics in the classroom*. London: Routledge.
Hudson, T. (1983). Correspondence and numerical differences between disjoint sets. *Child Development*, **54**, 84–90.
Hughes, M. (1981). Can pre-school children add and subtract? *Educational Psychology*, **1**, 207–219.

Hughes, M. (1982). Rappresentazione grafica spontanea del numero nei bambini: alcuni resultati preliminari. *Eta evolutiva*, **12**, 5–10.

Hughes, M. (1983). What is difficult about learning arithmetic? In M. Donaldson, R. Grieve and C. Pratt (eds), *Early childhood development and education*. Oxford: Blackwell.

Hughes, M. (1986). *Children and number: Difficulties in learning mathematics*. Oxford: Blackwell.

Hughes, M. and Grieve, R. (1978). On asking children bizarre questions. *First Language*, **1**, 149–60.

Hurlock, E.B. (1964). *Child development*. New York: McGraw-Hill.

Hutt, E. (1986). *Teaching language-disordered children: A structured curriculum*. London: Edward Arnold.

Ifrah, G. (1987). *From one to zero: A Universal history of numbers*. Harmondsworth: Penguin.

Irvine, S.H. (1988). Constructing the intellect of the Shona: A taxonomic approach. In J.W. Berry, S.H. Irvine and E.B. Hunt (eds), *Indigenous cognition: Functioning in cultural context*. Dordrecht: Martinus Nijhoff.

James, N. and Mason, J. (1982). Towards recording. *Visible Language*, **16**, 249–58.

Jaworski, B. (1985). A poster lesson. *Mathematics Teaching*, **113**, 4–5.

Jaworski, B. (1988) 'Is' versus 'seeing as': Constructivism and the mathematics classroom. In D. Pimm (ed.), *Mathematics, teachers and children*. London: Hodder and Stoughton.

Joffe, L. (1981). *School mathematics and dyslexia*. Ph.D. thesis, University of Aston.

Joffe, L. (1983). School mathematics and dyslexia: A matter of verbal labelling, generalisation, horses and carts. *Cambridge Journal of Education*, **13**(3), 194–7.

Jones, P.L. (1982). Learning mathematics in a second language: A problem with more and less. *Educational Studies in Mathematics*, **13**, 269–88.

Kamii, C.K. (1985). *Young children reinvent arithmetic*. New York: Teachers College Press.

Kamii, M. (1980). Place value: Children's efforts to find a correspondence between digits and number of objects. Paper presented at the Tenth Annual Symposium of the Jean Piaget Society, Philadelphia, May.

Kamii, M. (1981). Children's ideas about written number. *Topics in Learning and Learning Disabilities*, **1**, 47–59.

Kearins, J. (1983). A quotient of awareness. *Education News*, **18**(4), 18–22.

Kearins, J. (1985). Cross-cultural misunderstandings in education. In J.B. Pride (ed.), *Cross-cultural encounters*, pp. 65–81. Melbourne: River Seine.

Kearins, J. (1989). Measurement and direction knowledge in Aboriginal and non-Aboriginal children of Western Australia. In D. Keats, D. Munro and L. Mann (eds), *Heterogeneity in cross-cultural psychology*. Lisse: Swets and Zeitlinger.

Kearins, J. and Butters, J. (1986). Cultural number learning and school expectation. Conference paper: 8th International Congress of Cross-Cultural Psychology, Istanbul, July.

Kintsch, W. (1986). Learning from text. *Cognition and Instruction*, **3**, 87–108.

Kintsch, W. and Greeno, J.G. (1985). Understanding and solving word arithmetic problems. *Psychological Review*, **92**, 109–129.

Klahr, D. (1984). Transition processes in quantitative development. In R.J. Sternberg (ed.), *Mechanisms of cognitive development*. New York: W.H. Freeman.

Klahr, D. and Wallace, J.G. (1976). *Cognitive development: An information processing view*. Hillsdale, N.J.: Lawrence Erlbaum Associates Inc.

Krutetskii, V.A. (1976). *The psychology of mathematical abilities in school children* (translated by Joan Keller). Chicago: University of Chicago Press.

Kuriyama, K. and Yoshida, H. (1987). Representational structure of numbers in children: Analyses of strategies in solving both addition and subtraction problems. *Bulletin of Miyazaki Women's College*, **14**, 13–20.

Kuriyama, K. and Yoshida, H. (1988). Representational structure of numbers in children. *The Japanese Journal of Psychology*, **59**, 287–94.

Labinowicz, E. (1985). *Learning from children: New beginnings for teaching numerical thinking*. Menlo Park, Calif.: Addison-Wesley.

Lakoff, G. and Johnson, M. (1980). *Metaphors we live by*. Chicago: University of Chicago Press.

Lancy, D. (1983). *Cross cultural studies in cognition and mathematics*. London: Academic Press.

Lee, R.B. (1968). What hunters do for a living. In R.B. Lee and I. De Vore (eds), *Man the hunter*. New York: Aldine.

Leonard, L.B. (1979). Language impairment in children. *Merrill-Palmer Quarterly*, **25**(3), 205–232.

Lewis, A.B. and Mayer, R.B. (1987). Students' miscomprehension of relational statements in arithmetic word problems. *Journal of Educational Psychology*, **79**, 363–71.

Light, P.H. (1986). Context, conservation and conversation. In P.H. Light and M. Richards (eds), *Children of social worlds*. Oxford: Polity.

Light, P.H., Buckingham, N. and Robbins, A.H. (1979). The conservation task as an interactional setting. *British Journal of Educational Psychology*, **49**, 304–310.

Ling, A.H. (1978). Basic number and mathematical concepts in young hearing-impaired children. *Volta Review*, **80**(1), 46–50.

Low, G.D. (1988). On teaching metaphor. *Applied Linguistics*, **9**(2), 125–47.

Lyons, J. (1977). *Semantics*, Vol II. Cambridge: Cambridge University Press.

McGarrigle, J. and Donaldson, M. (1975). Conservation accidents. *Cognition*, **3**, 341–50.

McIntosh, A. (1979). Some children and some multiplication. *Mathematics Teaching*, **87**, 14–15.

Madell, R. (1985). Children's natural processes. *Arithmetic Teacher*, **32**(7), 20–22.

Marshack, A. (1972). *The roots of civilization: The cognitive beginnings of man's first art, notation and symbol*. New York: McGraw-Hill.

Mason, J. (1987). What do symbols represent? In C. Janvier (ed.), *Problems of representation in the teaching and learning of mathematics*. Hillsdale, N.J.: Lawrence Erlbaum Associates Inc.

Mason, J., Burton, L. and Stacey, K. (1987). *Thinking mathematically*. London: Addison-Wesley.

Mason, J.M., Kniseley, E. and Kendall, J. (1979). Effects of polysemous words in sentence comprehension. *Reading Research Quarterly*, **15**, 49–65.

Meadow, K.P. (1980). *Deafness and child development*. London: Arnold.

Meek, C. and Feril, N. (1978). A pilot study into the level of some words taken from the Papuan New Guinea Community School Mathematics Syllabus. *Mathematics Education Centre Report No. 3*. Papua New Guinea University of Technology.

Menninger, K. (1969). *Number words and number symbols: A cultural history of numbers* (translated by P. Broneer. Cambridge, Mass.: MIT Press. (Translated from original publication, *Zahlwort und ziffer*, 1958, Vanderhoeck & Ruprecht.)

Messer, D.J. (1983). The redundancy between adult speech and non-verbal interaction: A contribution to acquisition. In R.M. Golinkoff (ed.), *The transition from prelinguistic to linguistic communication*. Hillsdale, N.J.: Lawrence Erlbaum Associates Inc.

Michie, S. (1984). Why preschoolers are reluctant to count spontaneously. *British Journal of Developmental Psychology*, **2**, 347–58.

Miles, T. (1983). *Dyslexia: The pattern of difficulties*. London: Granada.

Miles, T. and Ellis, N. (1981). A lexical coding deficiency. In G. Pavlidis and T. Miles (eds), *Dyslexia research and its application to education*. Chichester: John Wiley.

Miller, K. (1984). Child as measurer of all things: Measurement procedures and the development of quantitative concepts. In C. Sophian (ed.), *Origins of cognitive skills*. Hillsdale, N.J.: Lawrence Erlbaum Associates Inc.

Miller, K. and Stigler, J.W. (1987). Computing in Chinese: Cultural variation in a basic cognitive skill. *Cognitive Development*, **2**, 279–305.

Miller, P.H. and West, R.F. (1976). Perceptual supports for one-to-one correspondence in the conservation of number. *Journal of Experimental Child Psychology*, **21**, 417–24.

Miller, P.H., Heldmeyer, K.H. and Miller, S.A (1975). Facilitation of conservation of number in young children. *Developmental Psychology*, **11**, 253.

Miura, I.T. (1987). Mathematics achievement as a function of language. *Journal of Educational Psychology*, **79**, 79–82.

Miura, I.T. and Okamoto, Y. (1989). Comparisons of U.S. and Japanese first graders' cognitive representation of number and understanding place value. *Journal of Educational Psychology*, **81**, 109–113.

Miura, I.T., Kim, C.C., Chang, C. and Okamoto, Y. (1988). Effects of language characteristics on children's cognitive representation of number: Cross-national comparisons. *Child Development*, **59**, 1445–50.

Montague, M. and Bos, C.S. (1986). Verbal mathematical problem solving and learning disabilities: A review. *Focus on Learning Problems in Mathematics*, **8**(2), 7–21.

Morris, R.W. (1978). The role of language in learning mathematics. *Prospects*, **8**, 73–81.

Mugny, G., Perret-Clermont, A.N. and Doise, W. (1981). Interpersonal co-ordination and sociological differences in the construction of the intellect. In G. Stephenson and G.B. Davies (eds), *Applied social psychology*, Vol. 1. Chichester: John Wiley.

Murray, J. (1985). Maths and exploratory talk. *Mathematics in School*, **14**(4), 15.

Mussen, P.H. (1963). *The psychological development of the child*. Englewood Cliffs, N.J.: Prentice-Hall.

Nesher, P. (1988). Precursors of number in children: A linguistic perspective. In S. Strauss (ed.), *Ontogeny, phylogeny and historical development*. Norwood, N.J.: Ablex.

Nesher, P. (1989). The stereotyped nature of school word problems. *For the Learning of Mathematics*, **1**, 41–8.

Nesher, P. and Katriel, T. (1977). A semantic analysis of addition and subtraction problems in arithmetic. *Educational Studies in Mathematics*, **8**, 251–69.

Nesher, P. and Katriel, T. (1978). Two cognitive modes in arithmetic word problem solving. In E. Cohores-Fresenbory and I. Wachsmuth (eds), *Proceedings of the 2nd International Conference of the Group for the Psychology of Mathematics Education*, pp. 226–41, Osnabruck. July.

Nesher, P. and Teubal, E. (1975). Verbal cues as an interfering factor in verbal problem solving. *Educational Studies in Mathematics*, **6**, 41–51.

Nesher, P., Greeno, J.G. and Riley, M.S. (1982). The development of semantic categories for addition and subtraction. *Educational Studies in Mathematics*, **13**, 373–94.

Newell, A. and Simon, H.A. (1972). *Human problem solving*. Englewood Cliffs, N.J.: Prentice-Hall.

Newman, L. (1982). The roots of teaching and learning mathematics. *Directions (Gallaudet College)*, **3**(1), 37–9.

Newman, M.A. (1977). An analysis of sixth grade pupils, errors on written mathematical tasks. In M.A. Clements and J. Foyster (eds), *Research in mathematics education in Australia, 1977*, Vol. 1. Melbourne: Swinburne Press.

Newman, M.A. (1981). Comprehension of the language of mathematics. In J.P. Baxter and A.T. Larkins (eds), *Research in mathematics education in Australia, 1981*. Adelaide: Mathematics Education Research Group of Australia.

Nicholson, A.R. (1977). Mathematics and language. *Maths in School*, **6**, 5.

Nickerson, W., Zannettou, I. and Sutton, J. (1986). *Succeeding with the deaf student in college*, Special Needs Occasional Paper No. 4. London: Longman for the Further Education Unit.

Nolder, R. (1984). *Metaphor: Its influence on the teaching and learning of mathematics in the secondary school*. Unpublished M.A. dissertation, Chelsea College (now King's College), University of London.

Nolder, R. (1985). Complex numbers. *Mathematics Teaching*, **110**, 34.

Nolder, R. (1988). Metaphor: Its influence on the teaching and learning of mathematics in the secondary school. Unpublished M.A. dissertation, Chelsea College (now King's College), University of London.

Ogden, P.W. and Lipsett, S. (1982). *The silent garden*. New York: St Martin's Press.

Olson, A.L., Kieren, T.E. and Ludwig, S., (1987). Linking Logo, levels and language in mathematics. *Educational Studies in Mathematics*, **18**, 359–70.

Ortony, A. (ed.) (1979). *Metaphor and thought*. Cambridge: Cambridge University Press.

Panman, O. (1982). Homonymy and polysemy. *Lingua*, **58**, 105–136.

Pauwels, A. (1987). Empirische toetsing van het model van Riley, Greeno en Heller over de ontwikkeling van de probleemoplossingsvaadigheid bij eevoudige redactie-opgaven (Empirical testing of the Riley, Greeno and Heller model of children's skill in solving word problems). Unpublished Master's thesis. Leuven: Onderzoekscentrum voor Onderwijsleerprocessen, K.U. Leuven.

Perret-Clermont, A.N. (1980). *Social interaction and cognitive development in children*. London: Academic Press.

Peters, J. (1975). Language and mathematics teaching in the Open University. *Teaching at a Distance*, **2**, 31–4.

Piaget, J. (1952). *The child's conception of number*. London: Routledge and Kegan Paul.

Pimm, D. (1987). *Speaking mathematically: Communication in mathematics classrooms*. London: Routledge and Kegan Paul.

Pirie, S.E.B. (1988). Understanding: Instrumental relational, intuitive, constructed, formalised . . .? How can we know? *For the Learning of Mathematics*, **8**(3), 2–6.

Pirie, S.E.B. and Kieren, T. (1989). Recursion and the mathematical experience. In. L. Steffe (ed.), *The mathematical experience*. New York: Springer-Verlag.

Pirie, S.E.B. and Schwarzenberger, R.L.E. (1988). Mathematical discussion and mathematical understanding. *Educational Studies in Mathematics*, **19**, 459–70.

Pirie, S.E.B., Newman, C. and Schwarzenberger, R.L.E. (1989). Mathematical discussion: Incoherent exchanges and shared understandings. In C. Laborde (ed.), *Language and mathematics: Selected papers*. Coventry: University of Warwick.

Potter, M.C. and Levy, E.T. (1968). Spatial enumeration without counting. *Child Development*, **39**, 265–72.

Pratt, C. (1988). The child's conception of the conservation task. *British Journal of Developmental Psychology*, **6**, 157–68.

Preston, M. (1978). The language of early mathematical experience. *Mathematics in School*, **7**(4), 31–2.

Raven, J.C. (1947). *Progressive matrices*. London: H. Lewis/G. Harrap.

Resnick, L.B. (1983). A developmental theory of number understanding. In H.P. Ginsburg (ed.), *The development of mathematical thinking*. London: Academic Press.

Resnick, L.B. and Omanson, S.F. (1987). Learning to understand arithmetic. In R. Glaser (ed.), *Advances in instructional psychology*, Vol. 3. Hillsdale, N.J.: Lawrence Erlbaum Associates Inc.

Reynell, J. (1981). *Developmental language scales*, revised edn. Windsor: NFER-Nelson.

Riem, R. (1985). *Children learning to count: a social psychological reappraisal of cognitive theory*. Unpublished PhD dissertation, University of Kent at Canterbury.

Riley, M.S., Greeno, J.G. and Heller, J.I. (1983). Development of children's problem-solving ability in arithmetic. In H.P. Ginsburg (ed.), *The development of mathematical thinking*. London: Academic Press.

Robinson, W.P. (1984). The development of communicative conference with language in young children: A social psychological perspective. In H. Tajfel (ed.), *The social dimension*, Vol. 1. Cambridge: Cambridge University Press.

Rose, S. and Blank, M. (1974). The potency of context in children's cognition. *Child Development*, **45**, 499–502.

Ross, S.H. (1986). The development of children's place-value numeration concepts in grades two through five. Paper presented at the annual meeting of the American Educational Research Association, San Francisco, April.

Ross, S.H. (1988). The roles of cognitive development and instruction in children's acquisition of place-value numeration concepts. Paper presented at the annual meeting of the National Council of Teachers of Mathematics, Chicago, April.

Rudner, L.M. (1978). Using standard tests with the hearing-impaired: The problem of item bias. *Volta Review*, **80**(1), 31–40.

Rudnitsky, A.W. Drickamer, P. and Handy, R. (1981). Talking mathematics with children. *The Arithmetic Teacher*, **28**(8), 14–17.

Sainsbury, S. (1986). *Deaf worlds*. London: Hutchinson.

Samuel, J. and Bryant, P.E. (1984). Asking only one question in the conservation experiment. *Journal of Child Psychology and Psychiatry*, **25**, 315–18.

Sapir, E. (1921). *Language: An introduction to the study of speech*. New York: Harcourt Brace and World (reprinted, 1970. London: Hart-Davis).

Sastre, G. and Moreno, M. (1976). Représentations graphiques de la quantité. *Bulletin de Psychologie de l'Université de Paris*, **30**, 346–55.

Saxe, G.B. (1977). A developmental analysis of notational counting. *Child Development*, **48**, 1512–20.

Saxe, G.B. (1982). Culture and the development of numerical cognition: Studies among the Oksapmin of Papua, New Guinea. In C.J. Brainerd (ed.), *Progress in cognitive development research, Vol 1: Children's logical and mathematical cognition*. New York: Springer-Verlag.

Saxe, G.B., Guberman, S.R. and Gearhart, M. (1987). Social and developmental processes in children's understanding of number. *Monographs of the Society for Research in Child Development*, **52**,(2).

Schaeffer, B., Eggleston, V.H. and Scott, J.L. (1974). Number development in young children. *Cognitive Psychology*, **6**, 357–79.

Secada, W.G. (1985). Counting in sign: The number string, accuracy and use. *Dissertation Abstracts International*, **45**, 3571A (DER 85-02434).

Shiengold, K. (1987). The microcomputer as a symbolic medium. In R.D. Pea and K. Shiengold (eds), *Mirrors of minds: Patterns of experience in educational computing*, pp. 193–208. Norwood, N.J.: Ablex.

Shire, B. and Durkin, K. (1989). Junior school children's responses to conflict between the spatial and numerical meanings of 'up' and 'down'. *Educational Psychology*, **9**, 141–7.

Shuard, H and Rothery, A. (1984). *Children reading mathematics*. London: John Murray.

Shultz, T.R. and Horibe, F. (1974). Development of the appreciation of verbal jokes. *Developmental Psychology*, **10**, 13–20.

Siegler, R.S. and Robinson, M. (1982). The development of numerical understandings. In H.W. Reese and L.P. Lipsitt (eds), *Advances in child development and behavior*, Vol. 16. London: Academic Press.

Silverman-Dresner T. and Guilfoyle G.R. (1972). *Vocabulary norms for deaf children*. Washington: Alexander Graham Bell Association for the Deaf.

Simon, H.A. (1979). *Models of thought*. Newhaven, Conn.: Yale University Press.

Sinclair, A. (1988). La notation numerique chez l'enfant. In H. Sinclair (ed.), *La production de notations chez le jeune enfant*. Paris: Presses Universitaires de France.

Sinclair, A. and Sinclair, H. (1984). Preschool children's interpretation of written numbers. *Human Learning*, **3**, 173–84.

Sinclair, A. and Sinclair, H. (1986). Children's mastery of written numerals and the construction of basic number concepts. In J. Hiebert (ed.), *Conceptual and procedural knowledge: the case of mathematics*. Hillsdale, N.J.: Lawrence Erlbaum Associates Inc.

Sinclair, A., Siegrist, F. and Sinclair, H. (1983). Young children's ideas about the written number system. In D. Rogers and J. Sloboda (eds), *The acquisition of symbolic skills*. New York: Plenum Press.

Sinclair, H., Stambak, M., Lezine, I., Rayna, S. and Verba, M. (in press). *Infants and objects*. San Diego: Academic Press.

Skemp, R.R. (1979). *Intelligence, learning and action*. Chichester: John Wiley.

Snow, C. (1972). Mothers' speech to children learning language. *Child Development*, **43**, 549–65.

Snowling, M.J. (1987). *Dyslexia: A cognitive developmental perspective*. Oxford: Blackwell.

Solomon, Y. (1989). *The practice of mathematics*. London: Routledge.

Song, M. and Ginsburg, H.P. (1987). The development of informal and formal mathematical thinking in Korean and U.S. children. *Child Development*, **57**, 1286–96.

Sophian, C. (1987). Early developments in children's use of counting to solve quantitative problems. *Cognition and Instruction*, **4**, 61–90.

Souviney, R.J. (1983). Mathematical achievement, language and cognitive development: Classroom practices in Papua New Guinea. *Educational Studies in Mathematics*, **14**, 183–212.

Steffe, L.P. and Cobb, P. (1988). *Construction of arithmetical meanings and strategies*. New York: Springer-Verlag.

Steffe, L.P., von Glaserfeld, E., Richards, J. and Cobb, P. (1983). *Children's counting types: Philosophy, theory and application*. New York: Praeger.

Steinberg, R.M. (1984). A teaching experiment of the learning of addition and subtraction facts (Doctoral dissertation, University of Wisconsin-Madison, 1983). *Dissertation Abstracts International*, **44**, 3313A.

Stigler, J.W., Fuson, K.C., Ham, M. and Kim, M.S. (1986). An analysis of addition and subtraction word problems in American and Soviet elementary mathematics textbooks. *Cognition and Instruction*, **3**, 153–71.

Stone, P. (1980). Developing thinking skills in young hearing-impaired children. *Volta Review*, **82**(6), 345–52.

Struik, D.J. (1967). *A concise history of mathematics*. New York: Dover Press.

Stubbs, M. (1980). *Language and literacy*. London: Routledge and Kegan Paul.

Suppes, P. (1974). A survey of cognition in handicapped children. *Review of Educational Research*, **44**(2), 165–76.

Sutherland, R. (in press). Providing a computer-based framework for algebraic thinking. *Educational Studies in Mathematics*, **20**(3).

Sutherland, R., Hoyles, C. and Healy, L. (1989). *The role of pupil discussion in a computer environment*. Preliminary Report to the Leverhulme Trust.

Sutton, C. (1978). *Metaphorically speaking*. Occasional Paper. Leicester: Leicester University Press.

The Encyclopaedia Britannica (1963). Numerals, numeration, 610–612. Chicago: Encyclopaedia Britannica Inc. 31st edn.

Thomson, M.E. (1984). *Developmental dyslexia*. London: Edward Arnold.

Thomson, M.E. (1990). *Developmental dyslexia: Its nature, assessment and remediation*, 3rd edn. London: Whurr Publications.

Tirosh, A. and Graeber, A.O. (1989). Perservice elementary teachers' explicit beliefs about multiplication and division. *Educational Studies in Mathematics*, **20**, 79–96.

Tizard, B. and Hughes, M. (1984). *Young children learning*. London: Fontana.

Tomasello, M. (1987). Learning to use prepositions: A case study. *Journal of Child Language*, **14**, 79–98.

Tomlinson-Keasey, C. and Kelly, R.R. (1974). The development of thought processes in deaf children. *American Annals of the Deaf*, **119**(6), 693-700.

Tomm, K. (1989). Intentionality and consciousness in the work of Umberto Maturana. Public lecture, University of Alberta, February.

Tougher, H.E. (1981). Too many blanks! What workbooks don't teach. *Arithmetic Teacher*, **28**, 67.

Treffers, A. and Goffree, F. (1985). Rational analysis of realistic mathematics education: The Wiskobas program. In L. Streefland (ed.), *Proceedings of the Ninth International Conference for the Psychology of Mathematics Education, Vol. 2: Plenary addresses and invited papers*. Utrecht: Research Group on Mathematics Education and Educational Computer Center, Subfaculty of Mathematics, State University of Utrecht.

Tuma, D.T. and Reif, F. (eds) (1980). *Problem solving and education: Issues in teaching and research*. Hillsdale, N.J.: Lawrence Erlbaum Associates Inc.

Van Lehn, K. (1986). Arithmetic procedures are induced from examples. In J. Hiebert (ed.), *Conceptual and procedural knowledge: The case of mathematics*. Hillsdale, N.J.: Lawrence Erlbaum Associates Inc.

Verschaffel, L. (1984). Representatie-en oplossingsprocessen van eersteklassers bij aanvankelijke redactie-opgaven over optellen en aftrekken. Een theoretische en methodolgische bijdrage op basis van een longitudinale, kwalitatief-psychologische studie (First graders' representations and solutions processes on elementary addition and subtraction word problems. A theoretical and methodological contribution based on a longitudinal, qualitative-psychological investigation). Unpublished doctoral dissertation, Seminarie voor Pedagogische Psychologie, K.U. Leuven.

Verschaffel, L. and De Corte, E. (1990). Do non-semantic factors also influence the solution process of addition and subtraction word problems? In H. Mandl, E. De Corte, N. Bennett and H.F. Friedrick (eds), *Learning and instruction. European research in an international context. Vol. 22 Analysis of complex skills and complex knowledge domains.* (415–29). Oxford: Pergamon.

Vygotsky, L.S. (1962). *Thought and language*. Cambridge, Mass.: MIT Press/New York: John Wiley.

Vygotsky, L.S. (1978). *Mind and society: The development of higher psychological processes*. Cambridge, Mass.: Harvard University Press.

Wagner, S.H. and Walters, J. (1982). A longitudinal analysis of early number concepts: From numbers to number. In G.E. Foreman (ed.), *Action and thought: From sensorimotor schemes to symbolic operations*. London: Academic Press.

Walkerdine, V. (1988). *The mastery of reason*. London: Routledge.

Warden, J. (1981). Making space for doing and talking with groups in a primary classroom. In A. Floyd (ed.), *Developing mathematical thinking*. Milton Keynes: Open University Press/New York: Addison-Wesley.

Watson, I. (1980). Investigating errors of beginning mathematicians. *Educational Studies in Mathematics*, **11**, 319–30.

Webb, R.A., Oliveri, M.E. and O'Keeffe, L. (1974). Investigation of the meaning of different in the language of young children. *Child Development*, **45**, 984–91.

Werner, H. and Kaplan, B. (1963). *Symbol formation*. New York: John Wiley.

Whiteman, M. and Piesach, E. (1970). Perceptual and sensorimotor supports for conservation tasks. *Developmental Psychology*, **2**, 247–56.

Willis, G.B. and Fuson, K.C. (1988). Teaching children to use schematic drawings to solve addition and subtraction word problems. *Journal of Educational Psychology*, **80**, 192-201.

Wood, D. (1988). *How children think and learn*. Oxford: Blackwell.

Wood, D., Wood, H., Griffiths, A. and Howarth, I. (1986). *Teaching and talking to deaf children*. Chichester: John Wiley.

Woods, S.S. Resnick, L.B. and Groen, G.J. (1975). An experimental test of process models for subtraction. *Journal of Educational Psychology*, **67**, 17–21.

Yoshida, H. and Kuriyama, K. (1986). The numbers 1 to 5 in the development of children's number concepts. *Journal of Experimental Child Psychology*, **41**, 251–6.

Yoshida, H. and Kuriyama, K. (in prep.). The role of the number 5 in the development of children's number concepts.

Zaslavsky, C. (1973). *Africa counts: Number and pattern in African culture*. Boston: Prindle, Weber and Schmidt.

Zepp, R. (1989). *Language and mathematics education*. Hong Kong: API Press.

Zhu, J. (1989). Speed and rhythm of preschoolers' counting. *Proceedings of the Biennial Meeting of the Society for Research in Child Development*, **6**, 406 (abstract).

Name index

Subject index